Thomas Blizard Curling

Observations on the Diseases of the Rectum

Thomas Blizard Curling

Observations on the Diseases of the Rectum

ISBN/EAN: 9783337035136

Printed in Europe, USA, Canada, Australia, Japan

Cover: Foto ©berggeist007 / pixelio.de

More available books at **www.hansebooks.com**

OBSERVATIONS

ON THE

DISEASES OF THE RECTUM.

BY THE SAME AUTHOR.

A

PRACTICAL TREATISE

ON

THE DISEASES OF THE TESTIS

AND OF THE

SPERMATIC CORD AND SCROTUM.

With numerous Wood Engravings.

Third Edition. Revised and Enlarged. 8vo. Cloth. 16s.

OBSERVATIONS

ON THE

DISEASES OF THE RECTUM.

BY

T. B. CURLING, F.R.S.

CONSULTING SURGEON TO THE LONDON HOSPITAL,
LATE PRESIDENT OF THE ROYAL COLLEGE OF SURGEONS OF ENGLAND.

Fourth Edition, Revised and Enlarged.

PHILADELPHIA:
LINDSAY AND BLAKISTON.
1876.

PREFACE TO THE FOURTH EDITION.

In this new edition of my "Observations on Diseases of the Rectum," all the chapters have been carefully revised and a fresh one added. Such additions to our knowledge of the pathology and treatment of these diseases as I have been able to acquire during the last twelve years have been incorporated in the text.

Grosvenor Street, May, 1876.

CONTENTS.

CHAPTER I.

INTRODUCTORY OBSERVATIONS ON DISEASES OF THE RECTUM.

	PAGE
Persons most subject to these Diseases	1
Mode of examining, and making applications to the Rectum	2
Advantages of Anæsthetics in the Treatment of these Diseases	4
Operations on the Rectum	5

CHAPTER II.

IRRITABLE ULCER OF THE RECTUM.

Description of the Folds at the lower part of the Rectum	6
Seat and Characters of the Irritable Ulcer	ib.
Symptoms of the Disease	7
Persons most subject to it	9
Liability **to be** overlooked	ib.
Mode **of inspecting the Sore**	ib.
Treatment of **Irritable Ulcer**	10
Division **of the superficial fibres of** the Sphincter Ani	11
Treatment after the operation	12
Treatment by forcible dilatation	14
Sores and Ulcers at the margin of the Anus	ib.
Treatment	ib.

CHAPTER III.

IRRITABLE SPHINCTER MUSCLE.

	PAGE
Symptoms produced by an irritable Sphincter	15
Difficulty in Defecation	16
Treatment	17

CHAPTER IV.

NERVOUS AFFECTIONS OF THE RECTUM.

Irritable Rectum	19
Cases	ib.
Morbid sensibility of the Rectum	22
Characters of the complaint	ib.
Cases	23
Neuralgia of the Rectum	25
Cases	ib.
Obscurity of the causes	27
Affections chiefly mental	28

CHAPTER V.

HÆMORRHOIDS.

Arrangement of the Veins in the lower part of the Rectum	29
Different Forms of Hæmorrhoids	ib.
Structure of External Piles	ib.
Structure of Internal Piles	30
Appearances presented by Hæmorrhoids	32
Causes of Hæmorrhoids	34
Symptoms of External Piles	35
Symptoms of Internal Piles	36
Disorders of the Urinary Organs in connexion with this Disease	37
Influence of affections of the Uterus	38
Inflammation, Suppuration, and Sloughing of Hæmorrhoids	39
Bleeding from Piles, and periodical discharges of Blood from the Rectum	40

	PAGE
Case	41
The Sources of Hæmorrhage	42
Mode of examining Internal Hæmorrhoids	43
General Treatment of Hæmorrhoids	44
Treatment of External Hæmorrhoids	45
By Excision	46
Treatment of Internal Hæmorrhoids	47
Removal by Cauterization	50
Removal by Ligature	54
Management of External Piles after operations on Internal	56
Treatment after operation	57
Retention of Urine	58
Removal with the Ecraseur	59
Mode of arresting Hæmorrhage from Piles	60
Case of obstinate Bleeding Pile cured by operation	61
Comparison of the advantages of cautery and ligature	63
Use of Mechanical Aids	64

CHAPTER VI.

PROLAPSUS OF THE RECTUM.

Nature of Prolapsus	66
Pathological changes in Prolapsus	67
Causes of Prolapsus	68
Instances of immense Prolapsus in a very relaxed state of the Sphincter Muscle	69
Symptoms of Prolapsus	70
Strangulation of Prolapsed Bowel	ib.
Case with sloughing of Mucous Membrane	ib.
Troublesome bleeding in Prolapsus	71
Treatment in children	72
Treatment in adults	74
By Escharotics	75
By Operation	76
By Mechanical Support	77

CHAPTER VII.

POLYPUS OF THE RECTUM.

Forms of Polypus	78

	PAGE
Polypus in children	79
Treatment	ib.
Cases	80
Polypus in adults	82
Strangulation of Polypus	ib.
Case	83
Modes of removal of the fibrous Polypus	ib.

CHAPTER VIII.

VILLOUS TUMOUR OF THE RECTUM.

Structure of the Villous Tumour	85
Symptoms	ib.
Treatment	86

CHAPTER IX.

FISTULA IN ANO.

Description	86
Modes of origin	89
Phlegmonous Abscess near the Rectum	ib.
Ulcers in the Rectum	89
Situation of the inner opening	90
Fistula in phthisical subjects	91
Complicated Fistula	92
Symptoms of Fistula	93
Treatment of Rectal Abscesses	ib.
Mode of examining a Fistula	94
Treatment of Complete Fistula by Operation	95
Treatment after Operation	97
Treatment of Sinus passing high up	98
Treatment of Hæmorrhage after Operation	100
Cure of Fistula by Ligature	101
Fistula Trocar described	ib.
Case treated by Ligature	102
Treatment of Blind External Fistula	103
Treatment of Blind Internal Fistula	ib.

	PAGE
Case	104
Treatment of complicated Fistula	105
Operations for Fistula in phthisical subjects	106
Treatment of Fistula communicating with the Vagina	108

CHAPTER X.

CATARRH OF THE RECTUM.

Symptoms and Treatment	109

CHAPTER XI.

CHRONIC ULCERATION OF THE RECTUM.

Description of the Morbid Changes in Ulceration of the Rectum	110
Causes	111
Syphilitic Ulceration	112
Tubercular Ulceration	113
Symptoms of Chronic Ulceration	114
Treatment	115
Cases	ib.
Rodent Ulcer of the Anus and Rectum	120
Case	ib.

CHAPTER XII.

STRICTURE OF THE RECTUM.

Description of the Changes in the Coats of the Rectum occurring in Stricture	121
The Seat of Stricture	123
Double Stricture	ib.
Causes of Stricture	124
Frequency of Stricture in Women	ib.
Strictures produced by the Healing of Ulcers and Wounds	125
Cases	ib.
Period of life subject to Stricture	126

Local Symptoms of Stricture	129
Constitutional Symptoms	131
Mode of examining for a Stricture	132
Growths and Excrescences on the Mucous Membrane in Stricture	133
Causes of these Growths	134
Examination of Strictures high up the Rectum	135
Obstructions from Tumours external to the Rectum	136
Treatment by gradual Dilatation	137
Treatment by forcible Dilatation	138
Treatment by Incisions	139
Dangers attending Incisions	140
Treatment of Strictures high up in the Rectum	141
Dangers incurred in the Use of Instruments in these cases	ib.
General Treatment of Stricture	143
Local applications to the mucous surface	144
Hæmorrhage from the Bowel in Stricture	145
Mode in which Bougies operate in curing Stricture	ib.
Difficulty of curing Stricture of the Rectum	146
Cases of cure	147
Treatment of Intractable Stricture by Colotomy	151
Case	ib.

CHAPTER XIII.

CANCER OF THE RECTUM.

Morbid Changes in the Rectum produced by the different Forms of Carcinoma	154
Seat of the Disease	155
Symptoms	156
Age at which Cancer occurs	158
Pregnancy in Cancer of the Rectum	159
Frequency of Cancer of the Rectum in Men	ib.
Treatment	ib.
Excision of the Carcinomatous Rectum	160
Lumbar Colotomy for Relief of the Symptoms	162
Cases	ib.

CHAPTER XIV.

EPITHELIAL CANCER OF THE ANUS AND RECTUM.

	PAGE
Liability of the Anus to Epithelial Cancer	164
Cases treated by Excision	ib
Dangers from Hæmorrhage after Excision	170
Epithelial Cancer of the Rectum	171
Case	ib.
Effects of Epithelial Cancer, and Scirrhous, and Medullary Cancer on the Rectum, contrasted	173

CHAPTER XV.

MELANOTIC CANCER OF THE ANUS.

Case	174

CHAPTER XVI.

OBSTRUCTION OF THE RECTUM, AND OPERATIONS REQUIRED FOR THEIR RELIEF.

Obstructions at the termination of the Sigmoid Flexure from Stricture or Cancer	175
Symptoms indicating Obstructions in this situation	ib.
Mode of exploring the Rectum	176
Colotomy for the relief of the misery consequent on Stricture and Cancerous disease	178
Mode of performing Lumbar Colotomy	179
Bryant's Operation by Oblique Incision	181
Difficulties of the Operation	ib.
Conditions of an Anus in the Loin	183
Results of Operations for Colotomy by the Author	184

CHAPTER XVII.

ATONY OF THE RECTUM.

Defective Muscular Power in the Rectum	184
Case	185

	PAGE
Caused by free use of Enemata	185
Atony a cause of Fecal Accumulations	ib.
Treatment of impacted Fæces	186

CHAPTER XVIII.

ANAL TUMOURS AND EXCRESCENCES.

Fibrous Tumours	188
Warts	189
Degeneration of	190
ase	ib.

CHAPTER XIX.

ORGANIC CONTRACTIONS OF THE ANUS.

Contraction after Operations	190
Case	191
Contraction after Ulceration	ib.
Case	ib.

CHAPTER XX.

PRURIGO ANI.

Causes	192
Prurigo in affections of the Womb	193
Case	ib.
Treatment of Prurigo	194

CHAPTER XXI.

CONGENITAL IMPERFECTIONS OF THE ANUS AND RECTUM.

Classification	197
Case of Congenital Fecal Fistula	ib.
Relative frequency of the Forms of Imperforation	198
Causes of these Imperfections	ib.

	PAGE
Relations of the Peritoneum to the Bowel in Imperforation	200
Defects in the Pelvis	201
Imperforate Anus without deficiency of the Rectum	ib.
Treatment by Operation	ib.
Imperforate Anus, the Rectum being partially or wholly deficient	ib.
Treatment by Operation	202
Excision of Coccyx	203
Anus opening into a Cul-de-sac, the Rectum being partially or wholly deficient	204
Varieties of this Form	ib.
Treatment by Operation	205
Importance of securing the Bowel to the outer Wound	206
Case	ib.
Inconveniences of an Operation simply by Incision or Puncture	207
Unique Case operated on by Amussat	ib.
Imperforate Anus in the Male, the Rectum being partially or wholly deficient, the bowel communicating with the Urethra or Neck of the Bladder	208
Inconveniences of this Malformation	ib.
Treatment by Operation	209
Imperforate Anus in the Female, the Rectum being partially deficient, and communicating with the Vagina	210
Inconveniences of this Malformation	211
Operation for Enlargement of the Vaginal Outlet	212
Operation for establishing a new Passage at the natural site, and for closure of the Vaginal Outlet	213
Rizzoli's Operation for Detachment of the Rectal Outlet from the Vagina, and its Establishment in the natural site	215
Imperforate Anus, the Rectum being partially deficient, and opening externally in an Abnormal situation by a narrow outlet	216
Varieties in the two Sexes	ib.
Operation for Enlargement of the original Outlet	ib.
Cases	217
Formation of an Anus at the natural site	218
Cases of Operation on the Male	ib.
Cases of Operation on the Female	219

	PAGE
Narrowness of the Anus	222
Case	ib.
Treatment	ib.
Condition of the Anus established by Operation at the natural site	223
Causes of death in Imperforation	ib.
Causes of death after Operations	224
Changes in the Bowel consequent upon Partial Obstruction	225
Troubles in Defecation resulting therefrom	226

CHAPTER XXII.

COLOTOMY IN CONGENITAL IMPERFECTIONS OF THE ANUS AND RECTUM.

The Lumbar and Inguinal Operations considered	227
The comparative difficulties in performing these Operations	ib.
Author's Experiments and Observations on the Bodies of still-born Infants	228
The comparative dangers of these Operations	230
The comparative convenience of an Anus in the Loin, and of one in the Groin	231
Condition of the Anus in two successful Cases of Colotomy in the Groin	232
Case of successful Colotomy in the Loin in an Infant, and condition of the Anus at eight years of age	233
Condition of the Bowel beyond the artificial Anus some years after operation	235
Condition of the Anus in the Loin in a boy operated on by Amussat in infancy	236
Case of Left Inguinal Colotomy	237
Question of Operation in Right or Left Groin	ib.
Author's Experiments on the dead bodies of Infants	238
Boucart's Experiments	ib.
Bryant's Case of Inguinal Colotomy on right side	239
Mode of performing Inguinal Colotomy on Infants	240

ON DISEASES OF THE RECTUM.

CHAPTER I.

INTRODUCTORY OBSERVATIONS.

THE terminal portion of the alimentary canal—the rectum—is subject to numerous and varied derangements, dependent upon its structure, its peculiar office in the economy, and its relation to the important parts in its vicinity. As a class of diseases, those of the rectum are as common as any to which the human body is liable; and they give rise to sufferings, in many instances, not only severe, but often accompanied with depression of spirits, and an anxiety of mind, out of all proportion to the gravity of the disorders. Many of these diseases spring from habits prejudicial to health, engendered by sedentary pursuits, or consequent on indulgence in the luxuries of civilized life. They are, therefore, found to be most prevalent in the middle and upper classes of society. With few exceptions, there are no diseases which yield more readily and effectually to careful management and surgical treatment, or which in their results afford greater satisfaction to the practitioner.

Complaints of the rectum are liable to be mistaken for affections of the uterus, and even of the bladder and prostate. A discharge from a fistula in ano has been supposed to proceed from the vagina. Patients have been treated for obstinate diarrhœa, when the actual disease has been stricture in the lower bowel, or a lacerated perineum and sphincter, and obstructions referred to the abdominal intestines have been discovered when too late to exist in their pelvic termination. In the treatment, therefore, of diseases of this region, it is very important, that an early and careful examination should be made of the part affected. By neglect of this precaution, serious disorders, which, if detected in time, would yield easily to treatment, are allowed to make progress, and to become difficult of cure. In females, the delicacy of the sex too frequently leads to concealment of these complaints, and raises obstacles to an inspection of the seat of them. The chief information is to be gained by a tactile examination. We can discover in this way contractions in the passage, as well as tumours and excrescences; and by the practised finger ulcers may be detected, and their size and situation accurately ascertained. The examination should always be made with gentleness. This caution is especially required when the sphincter is irritable. Let the surgeon, in introducing his finger, well greased, proceed slowly, stopping at times until the sphincter becomes quiet and accustomed to its presence. The muscle will then yield, and allow the finger to pass on without pain. A rough attempt to penetrate excites resistance from the muscle and spasm, and the passage of the finger then occasions considerable suffering and after-distress.

The examination of the parts diseased, as well as the application of remedies, may, in many instances, be aided by the use of speculums. They are made of various kinds, some of them ill-adapted for the object in view. Thus, many of them are of little use in consequence of the bulgings of the mucous coat of the bowel between the narrow blades of the instrument; and, in others, the side opening is too narrow or small to admit of a sufficient exposure of the part diseased. It must be borne in mind that the parts requiring inspection are situated, with some rare exceptions, within an inch and a half of the orifice of the bowel, and generally within the circle of the sphincter.

There is no speculum which I have found so generally useful, both in the examination of the lower part of the rectum and in the application of remedies to this part, as a glass reflecting speculum with a large side opening and a conical extremity.

Fig 1, half-size.

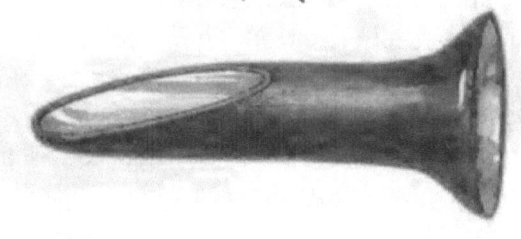

A small glass speculum with an open end made oblique is often very serviceable, especially in the application of caustic solutions to ulcers and strictures, but it cannot well be introduced if the sphincter be at all close or irritable. Before the use of any of these instruments the rectum should be emptied, and, if necessary, well cleansed by an injection of warm water.

Rectoscopes, instruments for obtaining a visual examination of the seat of disease high up in the bowel, have been contrived, but I have rarely succeeded in getting any useful information from them, or information unattainable by other means. With reflected light through a long glass tube I have recognized a congested state of the mucous lining of the upper part of the bowel, and have inspected the surface of syphilitic, dysenteric, and cancerous ulcers, and have been able in this way to make local applications to the diseased part.

In examining for mischief high up in the rectum, and hardly within reach of the finger, let the patient stand on the left leg with the right thigh and leg bent, the foot resting on a chair. Tell the patient to strain. This action and the weight of the abdominal organs will force the parts down, often low enough to admit of digital exploration. In some cases, when the sphincter is lax, a small hand, conically arranged and well greased, can be gradually insinuated into the rectum so as to explore the bowel high up, or to seize a projecting growth. In affections of the anterior wall of the rectum in females, this part may be readily everted by the pressure of one or two fingers passed into the vagina.

In the treatment of diseases of the rectum anæsthetics are a valuable auxiliary. In making examinations I have derived the greatest assistance and advantage from them. Under their influence the irritable sphincter relaxes, and a complete view can be had of the seat of disease in cases where pain and spasm would otherwise offer almost insuperable obstacles to a satisfactory exploration. And, in operations more painful than serious, anæsthesia has not only facilitated their performance, but saved the patient considerable suffering and distress.

It may seem superfluous to remark, that no operation, even of a trivial character, should be performed on the anus or rectum without due inquiry into the state of the patient's general health. I have heard of diffuse inflammation of a fatal character arising after the removal of a small excrescence from the anus, and after the division of a fistula; and of pyæmia occurring after the removal of hæmorrhoids; and although all operations are more or less liable to ill consequences, they very rarely happen after operations on the anus and rectum, except where the precaution alluded to is neglected. No prudent surgeon would undertake an operation in a person with a broken-down constitution, or with organic disease of the lungs or liver, or with albuminous urine; but with ordinary caution in the selection of cases, and with common care in performing the operations necessary, and in conducting the after-treatment, operations on the rectum are as successful and satisfactory as any belonging to surgery.

CHAPTER II.

IRRITABLE ULCER OF THE RECTUM.

The mucous membrane of the lower part of the rectum is arranged in longitudinal folds, which disappear in the expanded state of the bowel. These folds terminate below at the external sphincter. Just within this structure, and between the folds, the mucous membrane is slightly dilated, variously in different subjects, but in many to such an extent as to form small sacs or pouches. Besides these folds, and in the spaces between them, there is a series of short projecting columnar processes, about three-eighths of an inch in length, separated by furrows or sinuses, more or less deep, which are arranged around the lowest part of the rectum. In the evacuation of the rectum, foreign bodies or little masses of hardened fæces are liable to be caught or detained in the pouches just described. It is in these little sinuses, thus exposed to irritation, abrasion, and rent, that a superficial circumscribed ulcer is occasionally formed. On examining the ulcer, without distending the rectum, the lateral edges only being presented to view, the breach of surface has the appearance of a *fissure*—the term commonly given, but improperly, to this sore, which, though often originating in a rent, is obviously more than a mere cleft or fissure in the mucous membrane of the bowel. Such an ulcer may occur in any part of the lower circumference of the rectum, but it is very generally found at the back part, towards the coccyx. It is quite superficial, and, though sometimes circular, is generally of an oval shape, its long axis being longitudinal, and its lower extremity extending within the circle of the internal sphincter. On tactile examination, the breach in the mucous surface and the extent of the ulcer can be easily distinguished by a practised finger, especially when the edges are, as is often the case, somewhat indurated. With the speculum, the longitudinal folds being stretched out, the ulcer can be fully exposed, and it is then clearly seen not to be a mere fissure, but a superficial sore. The surface is of a brighter red than the surrounding membrane, and has the usual indented appearance of an ulcer. A small pedunculated pile or polypoid growth attached to the opposite side of the bowel is frequently found in these cases. The growth lodges in the ulcer, adding to the irritation and increasing the difficulty of cure.

The amount of suffering produced by this superficial ulcer varies a good deal, but the sore is generally extremely sensitive, and occasions severe distress. It is so situated that the fæces, in their passage outwards, rub over its surface, and the painful contact excites spasm of the sphincter muscle, causing a sharp burning pain, and often a forcing sensation, which lasts for two or three hours, the distress being usually greater after defecation than during the act, and in some instances an interval, varying from five minutes to ten or more, elapses between the evacuation and the occurrence of pain. The pain is sometimes so acute that patients resist the desire to pass their motions, and allow the bowels to become costive in dread of the sufferings brought on by evacuating them. I have also known patients to deprive themselves of food in order to avoid an action. In one case, the intensity of suffering led the patient, a young gentleman, to adopt the dangerous course of inhaling chloroform whilst sitting on the close stool, and he could not be persuaded to go to the closet without this remedy. The pain, though much increased during, and for some time after defecation, is in many cases constant—the patient never being free from a sharp lancinating pain, which disturbs rest, depresses the spirits, and renders the sufferer truly miserable. The least pressure at the anus gives uneasiness, so that the patient is obliged to avoid sitting, and either to rest on one hip or to lie down. He will sometimes place his finger on a spot outside the anus which exactly corresponds with the seat of the ulcer internally. The pains occasionally assume a neuralgic character, and are described as shooting up the back, down the limbs, or along the urethra. In the case of a man who suffered severely, the pain was referred to the spermatic cord and down the thigh on the same side, the pain in the cord especially being most distressing. The irritation may extend to the bladder, producing painful micturition, and even retention of urine. The stools are sometimes streaked with blood.

The sharp sufferings just described occur only when the sore is within the circle of the sphincter. I attended, with Dr. Arthur Farre, a lady who, after a hard evacuation, had a rent in the mucous membrane of the rectum just above the muscle. It could be felt with the finger, and plainly seen with a speculum. The ulcer which resulted caused much uneasiness after defecation, but it was much less persistent and distressing than the pain of the irritable ulcer, and the sore healed readily under local applications.

In comparison with many other diseases of the rectum, the irritable ulcer is not a common affection. The removal of hæmorrhoids, and the division of a fistula, may be performed with little risk of the sore consequent on the operation assuming the characters of the irritable ulcer. There are, however, exceptions. One of the most painful ulcers I have had to treat occurred, I was informed, after the excision of a small pile. In another case, in which I removed a large pile by ligature, the patient, a gentleman, neglected my injunction to keep at rest afterwards. He returned too soon to active business, and an irritable sore in the rectum was the consequence. I have also met with one which occurred after the removal of an internal pile by the acid nitrate of mercury in a lady of irritable constitution.

The irritable ulcer occurs usually in middle life, and is more frequent in women than in men. It is met with as often in single as in married women; and in persons of an hysterical temperament there are, occasionally, pains of so anomalous a character as sometimes to mislead the practitioner. Indeed, it is surprising how often this sore is overlooked even in common cases. Tactile examination is not always sufficient, for the sore is sometimes so superficial as not to be detected except by the most sensitive and practised finger. In all instances, therefore, of painful defecation for which the surgeon is unable to account, the rectum should be carefully examined with the speculum.

On the attempt to separate the margins of the anus, or to dilate the sphincter to get a view of the ulcer, or even to introduce the finger, spasm, with an aggravation of pain, is, in most cases, immediately excited; and the orifice becomes strongly contracted and forcibly drawn in. When this is the case, it is better to desist, and to get an assistant to administer an anæsthetic. As soon as the system is under its influence, the sphincter yields completely, and the surgeon is able to make a satisfactory exploration of the part, and to ascertain the exact seat, character, and extent of the ulcer. In cases free from spasm a good view may be obtained by gently dilating the anus with the two fore-fingers.

The irritable ulcer seldom heals under the influence of local applications. The treatment necessary is a longitudinal incision through its centre, including the superficial fibres of the sphincter muscle. The object of the operation is to place this muscle at rest for a time, and to enlarge the passage and displace the sore; thus removing those sources of irritation which prevent its healing. An incision, it is true, is not invariably required; but in all cases in which the pain is considerable, and in which there is much spasm of the sphincter, the attempt to procure the healing of the sore by local applications so often protracts the patient's sufferings, and so constantly ends in failure, that it is not desirable to make it.

The credit of originating the operation of incision for the cure of this painful complaint is due to the distinguished French surgeon Boyer. His operation was a free division of the sphincter muscle, a procedure unnecessarily severe.[1] Dupuytren practised

---

[1] A serious objection to the free division of the sphincter, an operation still practised by surgeons in this country, is a deficiency in the retentive power of the muscle, which sometimes distresses the patient after the part has healed.

a slighter **incision** than the operation performed by Boyer, and **the late Mr.** Copland **was content** to make **a simple** superficial incision of the part. In describing the operation to me, this excellent practical surgeon **spoke of** it as merely a division of the mucous membrane. I am convinced that on this point he is **in** error; at any rate this is not sufficient; and that however slight and superficial the incision may be, a few, at least, of the fibres of the sphincter must be divided. **I had occasion to** examine the rectum of a lady suffering from this affection, whilst she was under the influence of chloroform, and the parts being very lax, and in a good light, **I was able to** bring the ulcer well into view, and could distinctly perceive the fibres of the sphincter forming the bottom of the sore. Now, it is clear, that in such a case, or in an ulcer which has destroyed the mucous surface, an incision through the base of the sore must reach and divide muscular fibres.

In the evening **before the operation an aperient** should be given, in order that the bowels may **remain** at rest for two or three days after the incision. The patient should **be** placed on the left side, **with the** nates projecting a little over the edge of the bed, and the thighs bent, and opposite a good light. An anæsthetic can then be administered. The operation occupies so short a time that nitrous-oxyde gas is sufficient to produce insensibility. The division **of the ulcer may be performed in two ways; by an incision from within, or from without the rectum.** In **the latter mode, a** sharp-pointed bistoury is **carried** through **the base** of the ulcer, and the parts are divided by an incision from without inwards through the centre of the sore. A speculum must be previously introduced, to protect the opposite walls of the bowels from the point and edge of the bistoury. I prefer the operation from within outwards, which may be easily, and, indeed, more conveniently performed without the speculum. The ulcer being exposed and expanded by the two fore-fingers of an assistant, the cutting edge of a straight blunt-pointed bistoury is to be applied to the centre of the sore, which is to be divided by a slight superficial longitudinal incision. It is important to divide the fibres of the muscle at the extremity of the ulcer near the verge of the anus rather more freely than those above, so as to avoid any ridge or shelf on which the fæces would lodge. With this precaution the after-treatment becomes very simple, the parts being left very much to themselves. I have never been troubled with hæmorrhage, but if any vessel be seen pumping out blood, it may be seized and tied. I usually order a full dose of laudanum or *liquor opii* to be taken shortly after the operation, and to be repeated if necessary to keep the bowels quiet. In three or four days a mild aperient may be given, and repeated when required, to prevent costiveness and to render the motions somewhat soft. An injection of half-a-pint of warm olive oil given shortly before the first relief will soften the fæces and facilitate their passage.

The effect of the operation is remarkable. It at once relieves the severe symptoms, the pain experienced afterwards being merely the sore of the wound, and it rarely fails to secure the healing of the ulcer in the course of two or three weeks. The progress of the sore must, however, be watched until the surgeon is satisfied by an examination that the part is quite healed; for I have known of disappointment ensuing, and the painful symptoms returning, after the case had been given up under the supposition that the patient was cured. He should keep the recumbent posture. He need not remain in bed: rest on a couch is sufficient. If the healing of the ulcer proceed slowly, it may be touched with a camel's-hair pencil dipped in a solution of the nitrate of silver (gr. x. to ℥j.), or occasionally smeared over with some mild stimulating ointment. In an examination to ascertain if the sore be healed, it is better not to employ the speculum, which is liable to stretch the parts too much and to break the frail cicatrix. If a piece of wool, after being lodged in the part for a few hours, has no yellow stain; if no abrasion can be detected in a tactile examination, and the patient is quite free from soreness after stool, we may be sure the part is quite sound.

In cases of irritable sore, complicated with a pedunculated pile, the growth should be excised or tied after the incision of the ulcer.

Patients will not always submit to the division of the sphincter without a previous trial of other means; and where there is not much spasm, and but moderate suffering, the cure of the sore may often be obtained without it. The patient should remain at rest in the recumbent posture, and should take some mild aperient medicine to ensure soft evacuations. The ulcer may be brushed over occasionally with a solution of the nitrate of silver, or touched with the solid sulphate of copper, and a mild mercurial ointment smeared over its surface night and morning,—such as the *Unguentum hydrargyri*, or the *Unguentum hydrargyri nitratis*, diluted with two parts of lard. For sensitive ulcers, ointments containing the extract of belladonna are recommended; and this drug is sometimes added to mercurial applications. It may be used in the proportion of from ℨj.—ℨij. to ℥j. of lard.

In France another mode of treating the irritable ulcer is resorted to, viz. forcible dilatation and rupture of the sphincter muscle by the thumbs introduced at the anus. This proceeding was originally suggested by Recamier, and it is said to possess the great advantage of not requiring the patient to remain at rest in bed beyond a few hours, after which he can follow his usual occupations. It is admitted, however, that the treatment by dilatation is less sure than by incision. Giraldé mentions three cases of women in whom he found dilatation fail, and he was obliged to have recourse to incision.[2] The operation of incision is so simple, so effectual, and so harmless, that I see no reason to substitute for it a proceeding so rough and so uncertain as forcible dilatation.

The delicate skin at the margin of the anus is subject to a linear abrasion or chap, and small sores occasionally form between the folds of integument at the outer edge of the sphincter, which probably originate in an affection of the follicles of the part. These chaps and sores cause more or less uneasiness in defecation, and often give rise to troublesome itching; but they are never attended with spasm of the sphincter or with the severe pain which occurs in ulcer of the rectum, and there is seldom any difficulty in getting them to heal. The daily operation of cleansing should be performed with sponge and water, and a piece of

---

[2] "Dictionnaire de Méd. et de Chirurg. Pratiques," vol. ii. p. 681.

soft linen. All rough treatment of the part should be carefully avoided. If excoriations merely exist, the surface may be dusted occasionally with hair powder. When there are chaps or sores, a piece of cotton wool soaked in black wash, or in a lotion of the oxide of zinc (℥j. to ℨvj.), or in the common Goulard, and lodged in the part, is generally sufficient to cure them. The small sores sometimes require to be touched with the sulphate of copper. A lotion of the nitrate of silver is objectionable, as it stains the linen. Both primary and secondary syphilitic sores occur in the immediate vicinity of the anus. In females, a chap or excoriation at this part may be readily inoculated by the matter from a sore on the vulva. A description of the appearances and treatment of such ulcers would be beyond the scope of this work, and I call attention to them, only, that if they should be met with, their true character may not be overlooked.

CHAPTER III.

IRRITABLE SPHINCTER MUSCLE.

Persons occasionally suffer from symptoms similar to those described in the last chapter, but more moderate in degree. There is pain in defecation, especially during solid motions, increasing afterwards, and lasting for half an hour or an hour. It is described as a forcing sensation, or a feeling as if the bowel were unrelieved. The anus is strongly contracted and drawn in by the action of the sphincter. Any attempt to examine the part induces spasm; and, the finger passed through it, is tightly grasped by the muscle, as if girt by a cord. In cases of old standing, the muscle becomes hypertrophied, and forms a mass, encircling the finger like a thick, unyielding ring. The spasm is not always confined to the anus, for the fibres of the internal sphincter occasionally become enlarged, contracting the lower part of the gut, and closing strongly around the finger. The fæces, however, are not streaked with blood; there is not the racking and constant pain experienced in the case of ulcer, and there are long intervals of ease and relief from pain, particularly when the patient's attention is engaged.

This irritability and hypertrophy of the sphincter sometimes produces serious trouble in defecation, owing to the expulsive powers of the bowel being insufficient to overcome the impediment caused by this muscle to the passage of the fæces. This was well marked in the case of a widow lady, aged forty-six, whom I saw with Sir George Burrows, the muscular walls of the rectum above the sphincter having been over distended and weakened by the free and long-continued use of injections. The same difficulty, and arising from the same cause, had long existed in a married lady, aged about forty, who was referred to me by a medical friend.

Irritability of the sphincter occurs commonly in hysterical females, or in nervous susceptible women who are accustomed to watch and to intensify every sensation. I have seldom met with it in other persons independently of some local source of irritation, as an ulcer or an inflamed internal pile; and I believe that in men simple irritability of the sphincter muscle is a rare complaint. The investigation of such a case is seldom satisfactory without an examination of the rectum with the speculum; and, in most instances of irritable sphincter, I am convinced that some direct cause of irritation may be discovered by this means. I attended an Italian who had severe spasm, consequent upon the irritation of an inflamed prostate from gonorrhœa. In women, the suffering attending this complaint is aggravated during the menstrual period.

The treatment required in this affection is mild laxatives, the local application of an ointment containing chloroform, opium, or belladonna, and the occasional passage of a bougie coated with the sedative ointment. In the passage of an ordinary bougie the circular fibres at the anus close tightly around it, and effectually wipe off the sedative ointment, so that very little, if any, reaches the mucous membrane lining the internal sphincter. To obviate this difficulty I have had bougies made conical in form, for easy passages, with shallow grooves in which some of the ointment is lodged, and in this way carried into the bowel. The bougie alone gives great relief in those cases in which an irritable sphincter offers resistance to the passage of the fæces, and it was used with great advantage in the two cases alluded to in the previous page. A tallow candle, the softest of bougies, is often sufficient for the purpose.

In obstinate cases, especially when the sphincter is hypertrophied, it sometimes becomes necessary to make a partial division of the muscle, which should be done on one side, towards the ischium. This usually succeeds in relieving the complaint, but in hysterical cases the benefit is not always permanent. In an unmarried girl upon whom I performed the operation, relief was experienced for a time; but some months afterwards she called on me again, complaining of her old symptoms. In these cases attention must be paid to the uterine functions. Steel medicines, the shower-bath, and sea-bathing, will be found beneficial. The complaint is somewhat capricious, so that what gives relief in one case or at one time, fails in another case or at another time, and after resisting our best remedies it sometimes subsides spontaneously.

When the spasm is so complete as to cause difficulty in defecation and obstinate constipation, a freer division of the sphincter may be required. Mr. Gowlland communicated two cases to the Hunterian Society,[s] in which he divided the muscle on both sides, considering the double operation to be necessary to give satisfactory relief. I have never had occasion to do this.

CHAPTER IV.

NERVOUS AFFECTIONS OF THE RECTUM.

THERE are few surgeons much engaged in practice who have not been consulted occasionally for some nervous affection of the rectum. The symptoms as well as the causes of the complaint are usually obscure, and the diagnosis not unfrequently perplexing. On analyzing the symptoms, they appear to consist, in some instances, in an irritability, or

[s] Proceedings, 1865.

too frequent an inclination to relieve the bowels; in others, in a morbid sensibility or undue tenderness of the part; and more rarely in an exaltation of sensibility independent of contact, constituting neuralgia.

1. IRRITABLE RECTUM.

In derangements of the alimentary canal, and of the organs connected with it, the fæces are often unhealthy and irritating to the mucous membrane; consequently, when passed into the rectum they excite uneasiness, with an urgent desire to void them. Pressing and painful calls are also experienced when the bowel is ulcerated and in other ways diseased. In "the irritable rectum" there is an inclination, more or less urgent, to empty the bowel, usually at inconvenient times, although the mucous membrane as well as the fæces are healthy, and often when there is little or nothing to expel. Such was the nature of the following case:—

Case 1.—The Rev. ——, aged sixty-four, a country rector, fond of literary pursuits, consulted me on October 2nd, 1857, on account of some troublesome affection of the rectum. He was a tall man, pale and feeble in appearance, but represented himself as being in good health. His habit was rather costive, and he strained usually in defecation. His complaint was an urgent desire to relieve the rectum at inconvenient times, in church, usually just before and during the performance of divine service, notwithstanding an effort in the closet had previously proved ineffectual. The desire often came on at prayer time, and left him when in the pulpit. He was subject to it also when attending public meetings, and occasionally whilst riding in a railway carriage. When an inclination was restrained by a violent effort, it passed away without any action of the bowels following. He had been subject to this affection about a **year,** and during this **period** he had been sometimes free from it altogether for **two or** three weeks. It was more liable to occur when he **was** under some anxiety. Some months previously he had consulted a physician, who advised him to clear the bowel by an injection of warm water on Saturday evening, and to take a sedative pill on Sunday morning, before service, and he had derived some benefit from this treatment. I made a careful digital examination of the rectum, and also passed a bougie, but could find no structural change to account for the symptoms. Viewing the case as one of "irritable **rectum**" connected with **an anxious state of** mind, **I prescribed an** aperient pill **to be** taken on Friday evenings, **to be** followed, if the **bowels** were not well relieved on Saturday, by a castor-oil injection in the evening, and **a suppository** of soap and opium to be passed on Sunday **morning, and at other** times when he was likely **to be troubled.** The dose of opium in the suppository was to be **so regulated as to prevent drowsiness.** He was also urged to disregard the desire as much as possible. On November 6th following, my patient wrote to me from the country, stating that he had resumed his clerical duties, and had not been much troubled **since adopting my** suggestions.

Physicians have written **on** railway diseases. If a surgeon wrote **on such a subject he could scarcely overlook two complaints which have certainly become more common since the introduction of this** rapid **mode of travelling, viz. "irritable bladder" and "** irritable **rectum." The frequency of the** former may be inferred from the large sale of "railway conveniences," which are so graphically advertised; **and** no **one** who **has** witnessed the rush to the **recesses during a brief stoppage of an** express train **at a station on a cold day.** will be surprised that **nervous, fidgetty persons, and those labouring under** stricture and enlargement of the **prostate gland, who cannot ease themselves quickly, should be tormented by irritability of the** bladder. **Persons of** this **disposition are liable to suffer also** from irritability **of the rectum.**

Case 2. A gentleman, aged forty-five, enjoying **tolerable** health, and residing in the country, consulted me in September, 1855. He complained of seldom having a good and free evacuation, and of rarely **feeling** afterwards an adequate **sense of relief.** He stated that his fæces came away in lumps, and that he had the sensation on going to stool of an impediment existing in the passage. He had sometimes to go to **the closet twenty times a day,** voiding **only small scanty lumps.** He experienced **most discomfort from an** urgent desire to relieve the bowel when so situated as **to be** unable to resort to the closet, as when travelling by **railway.** He had been subject to this complaint many years, and it was increasing upon him. Purgatives rather aggravated it. On **examination** I **could find no** structural change, nor tenderness **within the rectum, and no irritability of** the sphincter. I **prescribed some chalk mixture, with sulphate of magnesia and cubeb powder, to be taken three times a day, and advised** his giving **himself a warm-water injection daily.** He returned into the **country, and I have since had no account of him.**

Case 3.—A gentleman, about sixty years of age, **of** an anxious temperament and most punctual habits, engaged in business in the city, and residing twenty miles out of London, about a mile from a railway station, after the usual relief at home in the morning, **was constantly teased just as he was leaving his house, or after his arrival at the station to meet the train, by a feeling of his rectum being insufficiently emptied. If time admitted of** his again going to the closet, he **would void** usually **one or two** lumps, after which the disagreeable sensation would **cease,** otherwise he was annoyed by it during his journey **to town.** On getting to his counting-house and to his occupations, the desire usually passed off without any action of the bowel. **This gentleman derived relief from clearing the rectum by an injection of warm water just before starting. When the complaint first troubled him there was no closet at the station, but since one has been put up he has been much less subject to it.**

The circumstance last mentioned shows how much this affection is dependent on an anxious, fidgety state of mind, against which patients may often successfully struggle.

2. MORBID SENSIBILITY OF THE RECTUM.

Several cases have fallen under my notice in which uneasiness has been experienced at a particular spot in the rectum, being complained of, chiefly, during, or after, defecation. The fixity, and, sometimes, severity of the pain, and its aggravation from pressure, have naturally led to the suspicion of the existence of some lesion in the mucous membrane, such as an ulcer; but on careful examination no breach of surface has been discovered, nothing, except in some instances, slight elevations and increased redness, and vascularity at the spot affected. The complaint consists chiefly in an exalted sensibility of the nerves of the part, but the alterations in appearance just mentioned indicate that there is also some slight superficial structural change. In a few instances I have found the morbid sensibility to occur after some previous affection which had been cured by an operation. The remedies for the complaint are chiefly local. Sedatives, such as opium and belladonna, introduced into the rectum, often give relief, but more permanent benefit may be derived from applications calculated to alter the character of the part, such as the solid sulphate of copper, or a solution of the nitrate of silver, applied to the spot through a speculum, and mercurial ointments.

Case 4.—In the spring of 1857 I saw, in **consultation** with Dr. Randall, of Portman-street, a French gentleman, engaged in mercantile business, a tall, healthy-looking man, aged twenty-six, on account of a painful affection of the rectum, to which he had been subject for two years. He complained of being troubled with a heat, or burning sensation in the bowel, experienced always after defecation. He suffered from it also at other times, indeed he was seldom free from uneasiness. The sensation was increased after taking wine. It first occurred after an attack of gonorrhœa, for the cure of which he had taken largely of copaiba capsules with iron. He had consulted several eminent surgeons in Paris, but had not been able to get relief. The seat of uneasiness was a spot about an inch within the anus. On careful examination I could discover no lesion of the mucous membrane, only two small papillated eminences. To these I applied the solid sulphate of copper, and advised the daily application of the mild ointment of the nitrate of mercury, and his taking a gentle aperient of confection of senna with sulphur. Dr. Randall afterwards informed me that our patient pursued this treatment for two days with the desired benefit, when he was suddenly summoned to Bordeaux, since which he had not been heard of.

Case 5.—In 1857, a gentleman about sixty-five years of age, in good general health, leading an active professional life, consulted me respecting a troublesome pain in the rectum, to which he had been subject about two years. It appeared that some years previously he had suffered from piles, and had undergone two operations for their removal, since which he had been free from the hæmorrhoidal complaint. His bowels were regular, and he passed no blood at stool. The pain complained of was described as of a dull, aching character, and, though not severe, proved very annoying. It occurred generally after a relief from the bowels, lasting an hour or two, but it troubled him also at other times, varying in duration and degree on different days. He had been repeatedly examined for the discovery of the cause, and about a year previously, under the supposition that the pain was dependent on an internal fissure or ulcer, an incision was made into the mucous membrane, but without affording any relief. The pain was referred to one particular spot a short distance within the rectum on the left side, and on examination with the finger, pressure at this spot caused the usual pain, and left a dull sense of uneasiness there afterwards. On introducing a glass speculum, I observed a small red patch in the mucous membrane at the part corresponding to the seat of pain. The patch did not resemble an ulcer. There was no breach of surface, and when the part was touched or rubbed no bleeding ensued. The spot was slightly elevated, and of a redder colour than the surrounding mucous membrane. I touched it with the solid sulphate of copper, and after an interval of three days applied a strong solution of the nitrate of silver (two scruples to the ounce). As benefit appeared to be derived from the treatment, the latter application was repeated four times, at intervals of three or four days. My patient then left London for his autumnal holidays, and on his return at the end of two months reported that he had experienced considerable relief, and had not been at all annoyed by the pain, though he was still conscious of a tender spot within the rectum. Two years afterwards, he informed me that he was still reminded occasionally of his complaint, but that he suffered no material annoyance from it.

I could add to these cases of morbid sensibility of the rectum, having preserved brief notes of some others very similar, but the two related are sufficient to show the character of the symptoms and the treatment calculated to relieve them.

3. NEURALGIA OF THE RECTUM.

The two forms of nervous affection already described would be included by some writers under the general term of *neuralgia*, the sensibility of the rectum being in a measure perverted or augmented; but it will be remarked, that in the first no actual pain is experienced—there is merely an irregular and often causeless desire to evacuate the part; while in the second, the uneasiness consequent upon the augmented sensibility is either produced or aggravated by friction and pressure. In true neuralgia of the rectum the pain is severe, but quite independent of contact. There is no tenderness.

Case 6.—In May, 1855, Mr. B——, a tall, strong, healthy-looking man, aged thirty-six, a lawyer, of abstemious habits, consulted me on account of a most troublesome and annoying affection of the rectum. He complained of suffering from a dead, aching pain, high up on the right side of the bowel. The pain was constant, but varied a good deal in intensity, and it was often so severe as to disturb his rest at night. He always referred the pain to the same spot, either externally deeply seated in the pudic region, or high up inside the rectum. His bowels were regular, and his complaint was not aggravated by their action. He had been a sufferer in this way for more than two years, during which time he had been under the care of a hospital surgeon of considerable experience, who, after trying various remedies, at length advised the division of the sphincter muscle. On examination, I found the anal aperture healthy and the sphincter free from irritability, and on using a speculum, I could find no ulcer or sore, so that I gave no encouragement to the operation. A full-sized bougie passed a considerable distance without difficulty, and without causing pain on reaching the upper part of the rectum. He had made excursions into the country, and had travelled on the

Continent, without any permanent beneficial result. His suffering was only slightly diminished during active walking exercise. The only remedy from which he had derived any relief was the injection of a solution of the extract of opium. On using this in the morning after an action of the bowels, the pain was lulled, and he was able to attend to business. Without this remedy, the pain was so severe as to make him uncomfortable, restless, and unfit for professional occupation. Finding that he derived slight relief from the pressure of a bougie, I advised his wearing a long plug, but as this did not reach the part affected, it was of no service. I tried successively the carbonate of iron in large doses, the liquor arsenicalis, and the oleum terebinthinæ, but without any good result. Quinine and steel he had already taken with no benefit. The tincture of aconite was also taken without effect. It was suggested that the pain might be dependent on an exostosis in the pelvis pressing on a nerve. Nothing of the kind could be detected, but he was advised to take the iodide of potassium on the chance of getting relief. This gentleman, after remaining under my care some months, ceased his attendance, and I lost sight of him until after seven years he wrote to inform me that his sad trouble still continued as bad as ever, the pain and the effects of the opium rendering life quite a burden to him.

Case 7.—An officer in the army, aged twenty-four, on duty at Chatham, a stout, muscular, healthy-looking man, consulted me in October, 1857, on account of a burning sensation in the rectum, which had troubled him more or less for three years. He referred his uneasiness to no particular spot, except that it was confined quite to the lower part of the bowel. It did not seem connected with the functions of the part, but it troubled him almost daily, coming on at uncertain times. He was occasionally free from uneasiness for a whole day. The complaint was aggravated by drinking wine or spirits. On repeated careful examinations I could find no piles, nor fissure, nor any lesion to account for the symptoms, and there was no irritability of the sphincter muscle. He had been under the care of several surgeons, but no treatment had afforded him the least relief. I gave directions for the regulation of his diet and bowels, and recommended the local application of mild citrine ointment with extract of belladonna; but no benefit resulted. I subsequently prescribed quinine and steel pills, and the use of a lotion of oxyde of zinc and dilute hydrocyanic acid. This treatment afforded him considerable relief. He left Chatham on leave, and spent two months in Ireland. On his return at the end of December he was not quite so well, and stated that the lotion had lost its influence. I made further alterations in the treatment, but the patient shortly ceased his attendance, being ordered to India.

In these two last cases the pain was not characterized by paroxysms, or by a suddenness of attack and disappearance, nor by any regular intermittence such as is often witnessed in neuralgia; nor was the uneasy sensation of an acute kind; but it was a continuous, enduring pain, sufficiently severe to interfere seriously with the comforts and even the business of life. It must be observed, too, that the pain was in no degree mental; for the patients were not persons of an anxious, nervous temperament, and, unlike some of the former cases of nervous affection, occupation and amusement had little influence in mitigating their troubles. In neither of the cases could I discover anything which could shed the faintest light upon the cause of the neuralgia. The remedies calculated to give relief are such as are useful in neuralgia elsewhere, as quinine, steel, arsenic, bromide of potassium, local sedatives and hypodermic injections, and they are as uncertain in removing the affection of the rectum as in curing neuralgia of other parts.

The division of these nervous affections into three

classes seems fully warranted by the cases which have been just related, and the distinction is calculated to prove a very useful guide to treatment. But in some instances it is difficult to refer nervous complaints of the rectum to either group, morbid sensibility and neuralgia being so combined as to prevent any strict classification.

To these forms of nervous affection I might add another which seems to be chiefly mental. It is well known that attention concentrated on a part may render it the seat of morbid and painful sensations. Thus a woman who has lost a parent or friend from cancer of the breast, will sometimes fix her attention so intently on this organ as to imagine it to be the seat of a tumour with lancinating pains, and patients will occasionally also so exaggerate sensations in the rectum until they believe the part to be affected with some formidable organic disease. In some instances there is a foundation for anxiety, in the existence of some slight curable complaint, and in such a case I know of no duty more pleasing to the surgeon than the removal of the cause, and with it the mental apprehension. A gentleman from Cheshire entered my room, looking so sallow, careworn, and miserable, that I quite expected to find him labouring under some serious organic disease. For six months he had been suffering from painful defecation, and after brooding over his trouble without consulting any one in his own neighbourhood, at length resolved to come up to London to ascertain his fate. I discovered a small ulcer just within the anus, which I incised superficially, and sent him home cured and a changed man in about a week.

CHAPTER V.

HÆMORRHOIDS.

THE hæmorrhoidal veins distributed in the submucous tissue at the lower part of the rectum communicate in loops, and form a plexus which surrounds the bowel just within the internal sphincter. The veins are best seen when somewhat congested, their deep purple hue being very apparent through the thin mucous membrane with which they are in close contact. The plexus is then seen to be about three-quarters of an inch in length, and composed of veins of various sizes, arranged for the most part lengthwise and in clusters, being especially collected in the longitudinal folds of the rectum. The plexus does not extend lower than the external sphincter, but branches from it, passing between the fibres of the internal sphincter, descend along the outer edge of the former muscle, close to the integuments surrounding the anus.

These hæmorrhoidal veins are very liable to become dilated and varicose, giving rise to the disease termed *hæmorrhoids* or *piles*. When the plexus beneath the mucous membrane within the external sphincter are thus affected, the hæmorrhoids are said to be *internal*: when the veins beneath the integuments outside the muscle are enlarged, the hæmorrhoids are called *external*. Both external and internal piles very frequently co-exist.

We may distinguish two kinds of external piles; 1. a sanguineous tumour; 2. a cutaneous excrescence or out-growth. The sanguineous tumour consists of a softish elevation of the skin near the margin of the anus, of a rounded form, and a livid or slightly blue tinge. On cutting into it, we find a dark-coloured coagulum enclosed in a cyst. It is open to question whether the tumour arises from dilatation of a hæmorrhoidal vein with thickening of its coat, or from rupture of the vessel, extravasation of blood, and the formation of accidental cyst around the clot. The rapid and even sudden formation of the tumour, and the impossibility in most cases of tracing any communication between the cyst and a vein, favours the conclusion that the pile originates in a rupture of the vessel. This kind of pile is generally single and seated at the side of the anus, but a second may form at a subsequent period.

The second form of external pile consists of flattened prolongations of skin from hypertrophy of the epidermis papillæ and cutaneous layers. They are generally the chronic results of the first form, a projecting fold left after absorption of the coagulum having undergone further growth. Like an elongated internal pile, the cutaneous excrescence seldom contains any element of the original disease, no clot and no enlarged or varicose veins, but clots and dilated veins may often be found at the base. There is often only a single broad, flat excrescence, situated at the side of the anus, but there are sometimes two, one on each side; and occasionally several, varying in shape and size, encircling the anus. Similar excrescences occur as the result of irritating discharges from the bowel, and are common in stricture and chronic ulceration of the rectum.

The changes in structure consequent upon internal hæmorrhoids vary a good deal. In general the lower veins of the plexus are dilated irregularly, or into pouches, which are filled with dark coagula. These coagula are often compact and hard. A section shows a number of veins of different sizes, mostly plugged with clots. A bunch of varicose veins, crowded in the lower ends of the longitudinal folds, produce prominent projections of the mucous membrane, and deepen the pouches between the folds. In addition to these elevations, a number of small dilated veins sometimes form in the short columnar projections described at page 6. Two or three of the larger prominences of the longitudinal folds meeting below coalesce, so as to form a transverse fold just within the sphincter. In the course of time the mucous membrane and submucous areolar tissue become more or less hypertrophied. Thus are developed elongated processes of a polypus form, which grow as much as one inch in length, and projecting transverse folds, measuring an inch and more in width. Not seldom there are two or even three transverse elevations of smaller size. The arteries, which are abundantly supplied to the lower part of the rectum, taking a longitudinal course towards the orifice, where they freely communicate, also enlarge considerably. The mucous membrane involved in internal piles is not only thickened, but extremely vascular. The disease is not confined to the smaller veins at the extremity of the rectum, but, as it makes progress, the larger veins higher up the rectum also become varicose.[4] Such are the changes

---

[4] Mr. H. Smith has described ("Holmes' System of Surgery") hæmorrhoidal tumours as occurring in two separate rows, one half an inch above the other, and he refers to some preparations in the Museum of the College of Surgeons as indicating this arrangement. The two rows in these preparations really consist of *external* and internal piles, so displayed in the bottles as to give the appearance of two rows. The statement is indeed a pathological error. I have never seen a double row of piles either in the living or dead subject.

found on examination of the rectum after death, but they convey only a faint and incomplete impression of the condition of the parts observed during life.

Internal piles seldom attract attention until they have become developed so as to protrude at the anus in defecation. They then exhibit a remarkable diversity of appearance, according to their number, size, and condition. The protrusion may consist of only one large pile, found usually towards the perineum. A single pile, consisting of a bright red projecting membrane connected with a loose fold of integument and readily extruded, often forms in young persons, especially women. More commonly there are three distinct prominent growths differing in size, one at each side of the anus, and a third in front, the latter, the perineal, being generally the largest. In old standing cases they may be more numerous, as many as four or five, but usually they do not exceed three. The distinction between them is commonly well marked, but not always, for the piles sometimes merge into each other, so that the protrusion forms nearly a circular prominence. The aspect of extruded piles depends much upon their condition, whether congested, inflamed, or constricted by the sphincter. In an inactive state, and in a relaxed condition of the sphincter, they form softish tumours of a red granular appearance, presenting just at the orifice of the anus; but when protruded and congested they constitute large tense tumid swellings of a deep red colour, and smooth surface, which readily bleed. When hæmorrhoids of large size are fully protruded, the integuments at the margin of the anus become everted, and form a broad band girting the base of the tumours externally. The skin thus everted, when swollen, irregular, and of a livid colour from deep congestion of the varicose veins, is liable to be mistaken for external piles, and, in consequence, is sometimes excised in operations, an error very likely to be followed by serious contraction of the anus.

External and internal piles often co-exist, and when this is the case the sphincter, lined by integument, commonly forms a narrow band separating the two. But it sometimes happens that the two forms merge into each other, the difference being recognized by the character of the covering, mucous membrane, or skin, the line of junction being visible on the surface of the tumours, as is well shown in the accompanying woodcut (1 1 1).

Fig. 2.

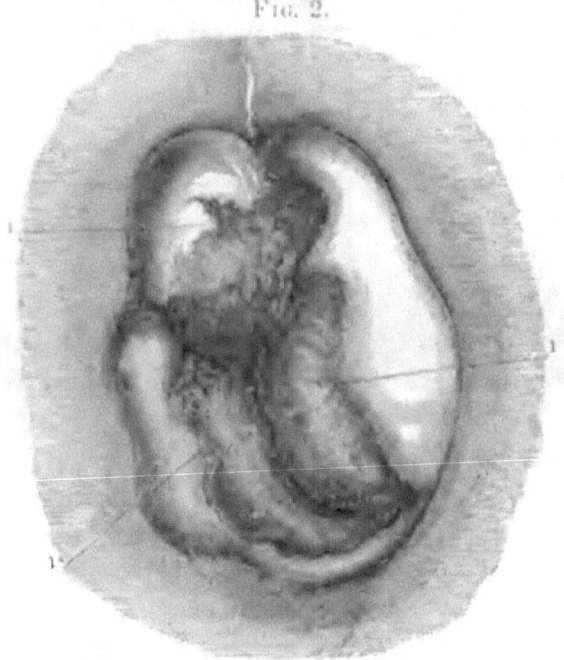

Hæmorrhoids is a disease of middle and advanced age. They rarely occur before puberty, and but few persons in after-life altogether escape them. All those circumstances which determine blood to the rectum, or which impede its return from the pelvis, tend to produce this disease. Drastic purgatives; the accumulation of fæces occurring in constipation; the strain on the coats of the veins taking place in protracted and forcible defecation, and in efforts to void the urine when the passage for it is obstructed; the impediments to the circulation caused in women by the gravid uterus, especially during labour, and by tumours of this organ, and in men by a greatly enlarged prostate gland; abdominal tumours pressing on the inferior mesenteric vein; disease of the liver interrupting the portal circulation, may all be regarded as causes of hæmorrhoids. There is, no doubt, in many persons a natural predisposition to the complaint, which is then produced by slight causes. This disposition is sometimes shown in a weak condition of the venous system generally. Thus, I have several times met with varicose veins of the lower extremities, and also varicose spermatic veins combined with hæmorrhoids. The disposition may be hereditary. The complaint, indeed, often occurs in members of the same family who inherit the local weakness of their parents. But a predisposition is more frequently acquired by sedentary habits, indulgences at table, and excitement of the sexual organs, which explains the well-known circumstance that hæmorrhoids are more prevalent in the higher classes of society than amongst the labouring population. The latter take plenty of exercise, live a good deal in the open air, and are little liable to constipated bowels. Hæmorrhoids, though a very common disease in both sexes, occur more frequently in males than in females. Few women, it is true, bear children without becoming in some degree affected by them; but the urinary and genital disorders of the other sex, combined with freer habits of living, are still more fertile sources of piles.

The symptoms produced both by external and internal piles vary a good deal in different subjects, and in different stages of the complaint. External piles cause a feeling of heat and tingling at the anus. A costive motion is followed by a burning sensation, and the excrescence becomes slightly swollen and tender on pressure, so as to render sitting uneasy. This congested state of the pile may pass off or lead to inflammation, accompanied with considerable enlargement of the hæmorrhoid, forming an oval tumour, red, tense, and extremely tender (fig. 3). The inflammation may subside, or go on to suppuration. When the matter is discharged, a clot of blood escapes with it, the

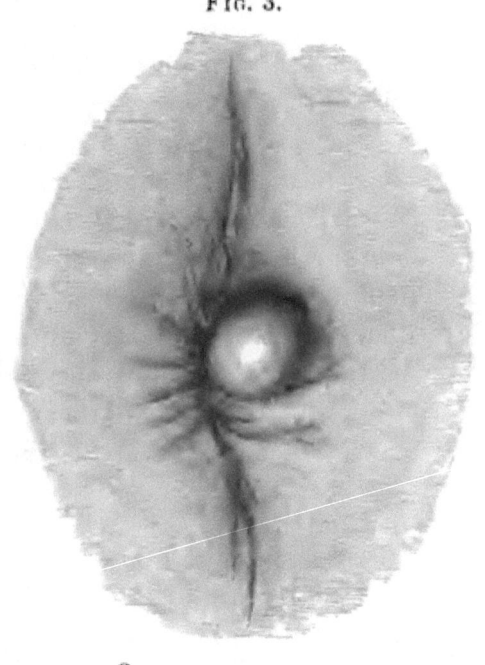

Fig. 3.

abscess closes, and the dilated vein is usually obliterated, the pile being reduced to a small flap of integument. Occasionally the opening remains fistulous. The irritation produced by costive evacuations, or by friction in sitting and cleansing the part, sometimes gives rise to ulceration on the inner surface of the pile, causing a sore, which extends a little within the circle of the sphincter. This is liable to occur particularly to those growths at the margin of the anus, which hold a middle place between internal and external piles. The pain in these cases is rather severe, a burning sensation lasting for an hour or two after defecation, and the sitting posture is at all times painful. The suffering, however, is not nearly so great as that occasioned by the irritable ulcer. External piles rarely give rise to bleeding to any great extent.

Internal piles, when slight, may exist for years, causing little inconvenience besides slight bleeding after a costive motion; and occasionally a feeling of fulness, heat, and itching, just within the anus. If only small, they protrude slightly with the mucous membrane in defecation, returning afterwards within the sphincter. When of larger size, the piles always protrude at stool, and require to be replaced, the patient usually pushing them up with his fingers. In a lax state of the sphincters, and in a loose and hypertrophied condition of the mucous membrane from which they spring, hæmorrhoids come down, even when the patient stands or walks about, so as to prove exceedingly troublesome, and to interfere with his taking walking exercise. When thus exposed to view they appear very prominent, of a rounded form, and often of a deep purple or violet hue, have a soft feel, and are evidently very vascular, bleeding readily when handled. If free from congestion, they exhibit a florid red colour with a rough granular surface. In consequence of the irritation from pressure and friction to which the protruding piles are liable, their mucous surface becomes tumid and abraded, and furnishes a free mucous discharge tinged with blood, which soils the linen. They are often so sore that the patient is obliged to keep the recumbent posture, the pressure in sitting causing great uneasiness. This is more particularly the case when the extremity of a projecting pile, forming a small rounded tumour, of a bright red granular aspect, constantly protrudes at the anus. A swelling of this kind is always more or less painful, and when inflamed or ulcerated is the seat of a sharp burning pain. Large piles within the sphincter, when swollen from irritation, sometimes occasion a sensation as if a foreign body were lodged in the part.

The symptoms produced by internal hæmorrhoids are not always confined to the seat of disease. Irritation frequently extends to the urinary organs, the patient being occasionally troubled with a frequent desire to pass water, and even with difficulty in voiding it, from spasm at the membranous part of the urethra. On the other hand, disease of the urinary organs is a very common cause of hæmorrhoids. The connexion, indeed, between piles and disorders of the urinary organs is a matter of considerable practical importance; and the surgeon should be careful to ascertain the original and chief source of the patient's sufferings. Persons with stricture in the urethra, stone in the bladder, or enlargement of the prostate gland, are accustomed to strain so much in passing water, that they are frequently unable to employ the bladder without at the same time relieving the rectum; and the disturbance in the circulation through the hæmorrhoidal veins produced in this way very often gives rise to piles. After the cure of the stricture in the urethra, or the removal of the stone from the bladder, the inconvenience suffered from the hæmorrhoids often ceases without any treatment particularly directed to the latter complaint. But the more frequent, and more severe and permanent complication is that of enlargement of the prostate, with hæmorrhoids; for not only are the hæmorrhoidal veins affected by the forcible efforts to relieve the bladder, but the enlarged gland, by obstructing the circulation in the larger veins, tends materially to promote the formation of piles. The veins of the hypertrophied prostate are always large and numerous, and communicate freely with the hæmorrhoidal; so that in congestion of the former the latter must more or less participate. Accordingly, few persons suffer from enlargement of the prostate without being also troubled with piles; and feeling a sense of weight and bearing down in the rectum, they are liable to attribute their symptoms to internal hæmorrhoids, instead of to the disease of the prostate gland.

Owing to the close relation of the uterus to the rectum, many of the diseases of the former organ have an injurious effect on the latter. In inflammatory affections of the uterus, the afflux of blood to this organ promotes the development of piles in the adjoining viscus. Tumours and diseases producing congestion of the womb also operate injuriously on the vessels of the rectum. Women usually suffer more from piles during the catamenia than at other periods, and if subject to bleeding, it occurs chiefly at the period of the menstrual flux. In some cases, the flow of blood from the rectum appears to be a compensation for a deficient discharge from the uterus.

Persons subject to piles frequently suffer no inconvenience from them until, irritated by an unusually costive motion, or by a smart purgative, or under the excitement of wine, the growths become congested and inflamed, and cause spasm of the sphincter muscle. They then have what is termed an "attack of piles,"—that is to say, they suddenly experience a sensation of heat, weight, and fulness, just within the rectum, followed by considerable pain at stool, and sometimes irritation about the bladder. These symptoms, which are often attended with febrile disturbance, arise from inflammation and swelling of the piles, which afterwards subside, but seldom without leaving some enlargement of the growths. The formation and increase of piles seem, indeed, to arise chiefly from a determination of blood to the rectum. This determination is greatly promoted by stimulating drinks, so that some patients never suffer from the complaint except after indulging in this way. They are then rendered sensible of an afflux of blood by a sense of heat or intolerable itching at the anus.

It has been stated that external piles are liable to inflame and suppurate, the matter forming a small abscess in the fold, which, bursting at its extremity, sometimes leaves a small fistulous opening. This gives rise to the discharge of a small quantity of pus, which appears as a dirty yellow stain on the linen, and leads the surgeon to suspect the existence of a **blind** internal fistula. On careful examination the **opening in the pile** may be discovered, and a fine probe passed into it goes to the bottom of what proves to be a blind sac, but which does not extend to the areolar tissue external to the rectum, and is not to be regarded and treated as a blind external fistula, the removal of the growth being sufficient for the cure of this kind of sinus. I once examined a specimen of **fistula in ano**, combined **with large internal hæmorrhoids**, which were riddled with **numerous minute holes, leading** to blind sinuses confined to the excrescences.

I have also remarked that when internal piles of some size protrude at the anus, they are liable to be constricted and strangulated by the external sphincter. The contracted muscle impedes the return of blood, and occasions inflammatory swelling of the piles, which may become strangulated and mortify. In this way hæmorrhoids of large size have been known to slough off, the patients being relieved of a serious complaint by a sort of natural process. An occurrence of this kind is attended with a good deal of pain and suffering, but is free from danger. In the cases which I have met with, the extremities only of one or two of the larger hæmorrhoidal growths perished, and the patients, though experiencing relief, were by no means cured of the disease.

One of the most common symptoms of internal hæmorrhoids, indeed that from which the name of the complaint is derived, is hæmorrhage, which occurs when the bowels are evacuated. The bleeding varies greatly in amount. Sometimes the motions are merely tinged with a few drops of blood: in other instances the quantity lost is considerable, several ounces being voided at stool. The bleeding may be irregular, occurring only after costive motions, or in certain states of health; or it may take place daily, going on even within the bowel, and producing the usual symptoms of derangement from continued losses of blood. Thus the complexion becomes blanched, and the lips appear waxy. The patient loses flesh and strength, has a quick and small pulse, suffers from throbbings in the temples, palpitations, and difficulty of breathing on making the slightest exertion, and at length finds his legs and feet swollen from œdema. The character of the bleeding also varies: it is sometimes venous, sometimes arterial. There are some persons who, without suffering any other inconvenience from a varicose state of the hæmorrhoidal veins, become liable to discharges of blood from the rectum, either at regular periods, or whenever, from good living or want of exercise, the habit is fuller than usual. In these cases from three to six ounces of blood, or even more, come away at stool, following the fæcal evacuation; and the blood which is voided is of a dark colour, and evidently venous. Such habitual hæmorrhoidal discharges are not uncommon in plethoric persons, as in the following instance:—

Case 8.—I had under my care, some years ago, a stout gentleman, upwards of seventy years of age, who had been subject to periodical discharges of blood from the bowels for many years, usually in the spring and autumn. After lasting a week or ten days, they generally ceased spontaneously, but not always; and when feeling faint and weak from their continuance, he was in the habit of arresting them by injections of cold water. The discharges at length ceased, but in six months afterwards his urine became albuminous; and, a year later, he died suddenly after an attack of epistaxis.

Periodical losses of blood from the hæmorrhoidal veins of this character relieve congestion of the liver and kidneys, help to ward off attacks of gout, and prevent fits of apoplexy. They are not, therefore, to be interfered with, unless, by their long continuance, they are exhausting the patient's powers. In many persons and states of constitution, and habits of life, they are rightly regarded as safety valves. These discharges, though hæmorrhoidal, cannot, perhaps, be strictly regarded as proceeding from hæmorrhoids, there being no change in the condition of the veins amounting to disease. It sometimes happens, however, that persons, after suffering from an attack of piles for a few days, have a pretty free discharge of blood from the rectum; the bleeding shortly ceases, and they find all their symptoms removed. This hæmorrhage is also venous. The escape of blood from the hæmorrhoidal veins unloads the congested and inflamed vessels, and thus the patient gets relief. But the bleeding which most commonly occurs from internal piles is undoubtedly arterial, taking place from arteries enlarged by the disease. The vessels on the spongy surface of the mucous membrane readily give way when blood is determined to the part in defecation, or when abraded by the passage of hard fæces. An artery of some size, exposed by ulceration, continues to pour out blood, weakening the patient, and giving rise to the symptoms above described. Sometimes a small artery on the prominent part of a protruded pile may be observed pumping out blood. The blood voided has a bright arterial colour. That hæmorrhage of this character is good for the health is quite a mistaken notion; and it is important that the practitioner should distinguish the bleeding taking place as a consequence of local disease, from that which arises from a constitutional plethora or congestion of the internal organs.

The only mode of making a satisfactory examination of internal piles is to obtain a thorough descent of the hæmorrhoidal growths. A tactile exploration is insufficient. It is difficult indeed to detect internal piles by passing the finger into the rectum, and it is often impossible to distinguish the soft flaccid hæmorrhoidal swellings from the loose folds of the mucous membrane in this way. When piles slip down easily, a straining effort by the patient will extrude them far enough for the surgeon's inspection. In other cases a lavement must be administered, and the expulsive effort used in voiding it should be kept up until the examination is made. By this means the full extent of the disease can be ascertained. I have already explained that in hæmorrhoids the mucous membrane from which they spring becomes relaxed and separated from the muscular coat, and following the movements of the piles gets extruded with them. The mucous membrane of the rectum may fall, however, into this condition independently of hæmorrhoids; and, in cases in which it descends freely, some discrimination is required to prevent the projecting folds from being mistaken for piles. The everted thickened membrane getting constricted by the sphincter becomes congested, and in this turgid and livid state is apt to be taken for swollen piles. I have witnessed this mistake, and know of an instance in which some folds were tied, under the supposition that they were hæmorrhoidal growths. In this case the parts constricted in the ligatures included the whole thickness of the intestine. Diffuse inflammation of the areolar tissue succeeded the operation, and had a fatal result.

When piles are small and cause but little inconvenience, the treatment is very simple. In all instances attention should be paid to the habits of living. Persons with this complaint should take wine in great moderation, if at all; and they are in most instances benefited by abstaining entirely from stimulating drinks. I have said that in the growth of piles there is commonly a determination of blood to the lower part of the rectum. Many individuals never experience a sense of this determination, or suffer from their piles, except after taking a glass of spirits and water or a few glasses of wine. Such persons should become rigid water-drinkers. Active exercise in the open air should be taken daily, and the patient must avoid sitting too long at the desk— I say at the desk, because it is by prolonged occupation in this way, and neglect of the rules of health, that hæmorrhoidal complaints are induced, which explains why literary persons so often suffer from them. Chairs with cane seats are to be recommended, as preventing the heat occasioned by stuffed cushions.[5] The most objectionable are those covered with the patent American cloth, which, being impermeable to moisture, causes a sensation of heat and closeness. The bowels must be carefully regulated, so as to

[5] Persons who are subject to piles, and take much carriage exercise, will experience comfort and advantage from using a moveable cane seat instead of the ordinary cushion

avoid hard and costive motions, as well as an undue action. Irritating the rectum by active and repeated purging is more hurtful even than constipation. On the other hand, when the secretions are sluggish and the bowels costive, a mild cathartic, by clearing the intestines, especially the large, unloads the congested vessels, and relieves the piles. In cases where the bowels are habitually costive, careful regimen, with sufficient exercise, will do much to correct the evil. But help from medicine is often needed. Linitive electuary, rendered more active if necessary by the addition of tartrate of potash taken at bedtime; or a daily dinner pill, consisting of the compound rhubarb, the extract of nux vomica, and the watery extract of aloes, will probably answer the purpose. The last preparation is not open to the objection commonly and justly made to the use of aloes in this complaint. The watery extract dissolves readily, and produces its effects before reaching the rectum. The foreign mineral waters, the *Püllna*, the *Friedrichshall*, or the *Hunyadi-János*, taken in the morning fasting, answer well with many patients, and ensure a comfortable relief. When the intestines require fully unloading, a draught, containing rhubarb powder and the tartrate or sulphate of potash, answers without producing local irritation.

In cases of external piles the parts should be sponged night and morning with cold water, or bathed with an astringent lotion of alum or sulphate of zinc. When they become inflamed, the patient must keep the recumbent position. The local application of pounded ice or of a freezing mixture will generally give complete relief in a few hours. If there be merely one pile of no great size, a rounded elevation of a livid colour (fig. 3), it is a good plan to open the swelling freely with a lancet, and then to squeeze out the dark coagulum. Let the patient remain in bed or on the sofa for two or three hours after the operation, and only a few drops of blood will be lost. If he sit up or move about, his dress may get saturated by the bleeding. The inflammation afterwards subsides, the vein becomes obliterated, and the pile shrivels up to a small fold. When several piles are affected, the common practice is to apply a few leeches, and to direct the parts to be well fomented and poulticed, and after the inflammation has subsided, to recommend the excision of the growths, to prevent the patient being again troubled with them. The application of ice is so effectual that leeching is seldom required. The excision of inflamed piles is a very painful operation; but since the introduction of anæsthetics, I have occasionally, in order to save time, removed them in this condition, the patient being placed under its influence. The bleeding after the operation relieves the inflammatory symptoms, and the part heals readily afterwards.

The excision of external piles is an easy operation, soon performed, and very effectual. The folds should be seized with the hæmorrhoidal forceps, drawn out a little, and then removed from the margin of the anus with a pair of knife-edged scissors curved on the flat. If a vessel be seen pumping out blood, it may readily be secured with a ligature. Gentle pressure by means of cotton wool lodged in the hollow and a T bandage will generally be sufficient to stop any bleeding. For an ulcerated pile excision is the best remedy. I removed from a married woman, aged thirty-seven, a patient in the London Hospital, a broad fold at the margin of the anus, the inner surface of which was the seat of rather a large superficial ulcer which extended a little within the sphincter. Though the fold was free from inflammation, she had severe pain for some time after going to stool, and had suffered in this way for seven months. The different practitioners to whom she had applied had treated her without making an examination. She was relieved at once by the operation, and in a fortnight the part was nearly healed, and she left the wards. This broad fold evidently belonged to the growths which hold an intermediate place between internal and external piles, the ulcer having formed on the mucous surface. The treatment applicable to external piles is proper for these mixed growths. They may be excised without any risk of troublesome bleeding. In the removal of these excrescences from the anus the surgeon should be careful not to excise the parts too extensively. I know of several cases, in which the skin at the base of some external piles having been freely cut away, the outlet has become so contracted afterwards as to cause much misery from difficult defecation. As the excision of large external piles is somewhat painful, recourse may be had to anæsthetics.

In cases of internal piles, half a pint of cold spring water thrown into the rectum in the morning after breakfast has a very beneficial effect on the hæmorrhoids by constringing the vessels and softening the motions before the usual evacuations. The relief afforded by this simple treatment, combined with care in the mode of living, is often remarkable. Persons who have suffered more or less from piles for years have assured me, that they have been quite free from all annoyance from them since they have regularly used the cold-water lavements. Some practitioners add alum, tannic acid, or muriated tincture of iron, to the water, to render the injection more astringent. In bad cases, I have used the decoction of oak bark with alum with much advantage. When an astringent injection is resorted to, it should be small in quantity, and given when the patient goes to bed, in order that it may be retained during the night, and thus have a longer time for acting on the piles. As an aperient, there is nothing better than the linitive electuary with sulphur, or the bitartrate of potass, which should be taken at bedtime, so as to ensure an action of the bowels in the morning. The confection of black pepper, better known as *Ward's Paste*, has long been in great repute as a remedy for piles, and there can be no doubt that it exerts a beneficial influence on the complaint. The usual dose is a drachm three times a day. This preparation is supposed to pass through the alimentary canal but little altered, and on reaching the rectum to act directly on the piles as a stimulating application. It does not seem a very scientific kind of practice to recommend patients to swallow a composition of pepper, which is to produce no effect until it reaches quite the other extremity of the alimentary canal; and Sir B. Brodie relates that a patient of Sir Everard Home taking this view of the matter, crammed as much as he could bear of it up the rectum, which, it is reported, had the effect of curing him. Sir E. Home afterwards used it as a local application in some other cases with manifest advantage. The cubebs pepper taken internally seems to relieve piles much in the same way as the confection of black pepper. I am not much in the practice of recommending these remedies, preferring to act more directly on the seat of disease. When there is a slight slimy discharge, and evidently an unhealthy state of the mucous surface of the hæmorrhoids, I find benefit derived from the application of a mild citrine ointment. The patient may take a little of this ointment on the end of his finger, and softening it at the fire, apply it to the parts within the sphincter every night. This is a better application than the gall ointment which is so often prescribed. An ointment of tannic acid is also beneficial. I have sometimes applied the solid sulphate of copper with good effect in correcting the granular condition of the surface of the piles; it is less painful than the nitrate of silver, which otherwise answers the same purpose. In cases where there is much irritation about the rectum great relief may be derived from the balsam of copaiba, which operates as a mild aperient at the same time that it allays irritation. It may be given in capsules, or in doses of half a drachm, with about fifteen minims of the liquor potassæ, three times a day, in a mixture to disguise the taste. Persons who find this mixture nauseous may be able to swallow the capsules.

When internal piles come down at stool, and require to be replaced, the patient should be provided in the closet with a basin of cold water and a piece of sponge, or soft linen rag, to apply to them. It may happen that in consequence of the protruded piles becoming a little inflamed, or more congested than usual, the patient finds himself unable to return them, and requires assistance. The surgeon should direct the patient to lie down on a sofa, and should endeavour by gentle pressure to empty the piles of blood, and then to push them back within the sphincter, in which he will generally be able to succeed if the hæmorrhoids have not been long down. They may, however, have become much swollen and congested, and be found tightly constricted by the sphincter. In this case, the piles should be punctured in several places with a needle, and afterwards bathed with cold or iced water, and the patient should be directed to remain in the recumbent posture. In a short time, the tension and swelling having subsided, the piles will very probably slip up without difficulty. If the protrusion have been strangulated for some time, and sloughing have already commenced, the surgeon ought not to interfere with them: fomentations and poultices should be applied, and attention must be paid to the general state of health, and suffering must be relieved, if necessary, by opiates.

Internal piles, when of such a size as to protrude at the anus, or when subject to inflammation, ulceration, and bleeding, so as to prove a constant source of annoyance and suffering, must be removed by operation. This may be done by cauterization, or by ligature. Excision was resorted to formerly, but dangerous bleeding so often ensued that this mode of getting rid of piles is not now practised.

Cauterization.—Dr. Houston, of Dublin, in a paper published in 1843,[6] strongly recommended the use of nitric acid for the cure of the florid vascular pile; and this escharotic has since been employed in numerous cases of the kind with an excellent result. The acid

---

[6] Dublin Journal of Medical Science, vol. xxiii.

nitrate of mercury is even more effectual, and is the escharotic to which I give the preference. It is certainly well adapted for destroying a bright fungous-looking growth, especially the single perineal pile so often seen in young people, and it is a safe and mild remedy. Means having been taken to bring the pile well into view, the patient should lean over a table, or lie on a couch on the side, and the nates should be separated by the hands of an assistant. The surgeon may then take a flat piece of wood, and, having dipped it in the acid nitrate of mercury, or in concentrated nitric acid, apply the escharotic to the entire surface of the hæmorrhoid, until its florid hue becomes quite changed to an ash colour. No speck of red should be allowed to remain. Care must be taken that none of the caustic fluid touches the skin at the margin of the anus. For the purpose of protecting the parts around the pile whilst making the application, I use a pair of steel forceps with electro-gilt blades, which are well adapted to grasp the base of the pile, and to shield the structures around. A spring catch or a screw in the handles will enable the forceps to act as a clamp. The moisture on the surface having been absorbed with lint, and the part smeared with sweet oil, the protrusion may be placed within the sphincter. The pain caused by the application in these slight cases is not severe, a mere burning sensation which soon passes off, and the separation of the superficial slough and healing of the sore occasioned by the

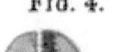

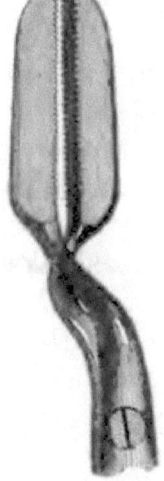

Fig. 4.

escharotic are attended with scarcely any uneasiness. If the pile be small, this plan answers very well, but it is not adapted for the removal of hæmorrhoidal flaps and tumours of any size. The treatment of piles by nitric acid has been unduly extolled, and this has led to its employment in very unfit cases. Indeed many instances in which the attempt has been made to remove well-developed growths by this method without success have come under my notice, and required other means of cure. In some instances, too, in which the nitric acid has been extensively applied, I have been informed that it produced considerable suffering, and that on the separation of the eschars there was troublesome hæmorrhage; and, also, that the large sores which ensued healed with difficulty. An officer in the army, who had lately returned from India with his health greatly impaired by constant hæmorrhage from piles, as well as by disease of the liver, came under my care. The late Sir Ranald Martin agreed with me that the operation by ligature would be attended with risk. I consequently applied the nitric acid pretty freely to a bunch of large internal hæmorrhoids. This was quite satisfactory in arresting the bleeding and preventing protrusion of the piles, but the sufferings produced were severe. The application was followed by inflammation and swelling of the external piles, and the patient kept his bed for a week. The actual cautery, Vienna paste, and other caustics, are also used, chiefly by French surgeons, and some ingenious instruments have been contrived for their application, especially by Amussat, who destroys internal hæmorrhoids by the application of Filho's solid caustic around the base of the pile.

Internal piles can be effectually removed by the actual cautery, and some excellent surgeons, English and German, prefer this mode of treatment. Before any operation, it is necessary to secure the complete protrusion of the piles. For this purpose a dose of castor oil should be given about six or eight hours before the time fixed for the operation; and, if the piles are not well in view, a pint of warm water should be thrown into the rectum on the surgeon's arrival. When the fluid is discharged the piles will descend, in which position they may be retained by a slight expulsive effort. The patient should lie on the side (I prefer the left), on a couch or bed of convenient height, with the thighs raised. The nates must be separated by an assistant. Ether can be given if the patient desires it.

The hæmorrhoidal growths are to be seized separately with the volsellum forceps and drawn out, and a firm clamp,[1] of the form represented on page 51, having been applied to the base and firmly closed, is to be held by an assistant. The projecting mass is then to be excised with scissors curved on the flat, leaving a good base, which is to be wiped dry. The cautery is to be applied to the divided surface, which is to be burned down to the level of the clamp. A cautery iron, or the galvanic cautery at red heat may be used for this purpose. After the destruction of the pile the clamp is to be taken off and applied to any others needing removal. The parts are afterwards to be well oiled and returned into the bowel with as little disturbance of the eschar as possible.

[1] Mr. H. Smith has made a valuable addition to the clamp of a layer of ivory to the outer surface, to prevent the transmission of heat to the parts with which it comes in contact.

In this operation Mr. H. Smith and Langenbeck of Berlin use the heated iron. I agree with Mr. Bryant and Esmarck of Kiel, in preferring the galvanic cautery as being less painful, and forming an eschar more secure against bleeding. It is a serious drawback, however, to the galvanic cautery that it requires a cumbersome apparatus not always to be had, especially in the country.

Fig. 5.   Fig. 6.

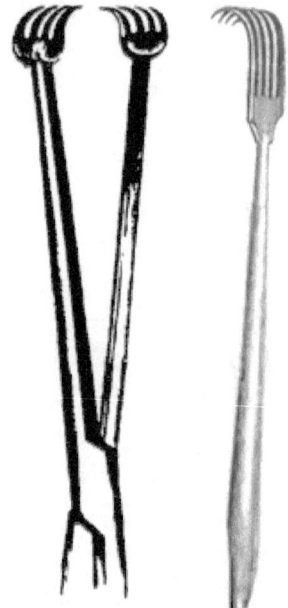

Ligature. — The instruments required for this operation are Salmon's deep four-pronged hook (fig. 6), or a doubled four-pronged forceps (fig. 5) having a spring catch to fix the grasp of the instrument, and a pair of strong, straight scissors with well-adjusted sharp blades, blunt at the points. A deeply-curved needle set in a handle with the eye near the point is sometimes required. I prefer and generally use the pronged forceps. Fine strong twine well waxed makes the best ligature.

Ether having been administered if required, and the buttock being drawn aside by an assistant, I seize and draw out a projecting pile, with the pronged forceps, and giving the instrument to the assistant make with the scissors a deep notch or incision along the base of the pile at the point of junction of the mucous membrane and skin, cutting parallel with the coats of the rectum and not into the pile, and carrying the ligature well into the groove, knot it tightly and securely at the root of the growth. When the hæmorrhoid is very broad or large in size, I pass the curved needle armed with a ligature to the bottom of the groove made by the incision, and through the central part of the base of the pile. The needle being withdrawn, the ligature is left double and tied on each side. Any remaining piles are afterwards to be treated in either of these ways. If there is no disposition to bleeding from the incised surfaces, which is seldom the case, the ends of the ligatures may be cut short, and the strangulated piles may be gently pushed up into the rectum. If hæmorrhage be apprehended, the ligatures can be left long, so as to enable the surgeon to draw the parts down and get at any bleeding vessel. In general, as soon as the parts are returned into the bowel all bleeding ceases. The free division of the outer base of the tumour greatly lessens the after-sufferings of the patient. The ligature in constricting the pile closes all the arteries leading into it, and effectually stops the hæmorrhage from the inner of the divided surfaces, the only one at all liable to bleed. The surgeon should be careful, however, to cut in the direction of the coats of the rectum, and not into the pile.

In operating on internal piles, it is not necessary to be particular to include in the ligature every portion of the morbid growth, or of the hypertrophied mucous membrane extruded with it. The removal of large piles leaves a sore surface of such an extent that the contraction which ensues in healing is sufficient to reduce any part that may have escaped the ligature, and to correct the lax condition of the adjoining mucous membrane which conduces to the protrusion of the hæmorrhoids.

If the surgeon when operating on internal piles should observe any of large size external to the sphincter, or any loose flaps of skin, he will do well to excise them at the same time. If the patient be not under the influence of ether, this is more felt than the tying of the ligature, but it will probably save some suffering afterwards, as the irritation produced by the ligatures is liable to cause the external piles to inflame.

Case 9.—Some years ago, a young clergyman in the country, who had suffered severely from piles, took lodgings in town, to undergo the treatment necessary for their cure. As he was troubled with both internal and external hæmorrhoids, I recommended the former to be tied, and the latter to be removed with the knife. Being a timid man, and finding that his sufferings proceeded almost entirely from the internal piles, he would only consent to my operating on these. The consequence was, that the irritation excited by the ligatures caused the external piles to inflame and swell excessively; and this added so much to his distress, that on the third day after the operation I was obliged to excise them. Fortunately, chloroform saved him from what would otherwise have been a most painful operation, in the inflamed state of the parts. He afterwards did quite well.

When the inflamed external piles are small, sufficient relief may be obtained by the application of ice. In excising these external growths, the surgeon should bear in mind the caution already given not to cut away healthy integument. In protruding hæmorrhoids of large size the everted skin at the margin of the anus is apt to be mistaken for external piles, and if this portion of skin be excised, contraction of the anus is very likely to ensue as the part heals. No inconvenient contraction ensues after the application of ligatures to internal piles alone.

I generally order a mixture with a full dose of opium to be given immediately after the operation, to be repeated in two hours if necessary, in order to relieve pain and bind the bowels. The sufferings of the patient afterwards vary a good deal, according to the extent of the parts strangulated and the irritability of the constitution, but they are generally slight and soon subside, a ligature on mucous membrane not being productive of as much pain as when constricting skin. In some instances, however, they are severe and prolonged, and accompanied with restlessness and want of sleep. When this is the case, there is nothing capable of giving such complete relief as ice. A bladder or piece of oiled silk, or, what is better still, a bag of vulcanized india-rubber, containing ice, may be applied to the part, and refilled as occasion requires. Both immediately after the operation and later, the greatest ease and comfort are derived from this application. Repeated doses of laudanum or morphia may be given if the pain continue. If the heat and swelling should be only slight, poultices and fomentations will give sufficient relief. No aperient should be taken for several days. I generally order a dose of castor oil on the fourth or fifth day after, and direct the patient to have a warm hip-bath prepared, and to drop into it directly after the medicine has acted. The bath is soothing, and gives great relief. An injection of half a pint of warm olive oil shortly before the expected relief will also help to lessen uneasiness. The tighter the ligatures are tied, the sooner they ulcerate through and come away. **The separation of the ligatures** usually occurs in about four **or five days, during which period the patient should remain in bed or on a couch.** The detachment of the sloughs leaves, of **course, at the** lower part of the **rectum a sore surface, and some attention** will be **required until this heals.** The motions must be kept soft by mild aperient medicine,—as the linitive electuary, castor oil, or mineral waters, and the patient should remain at rest chiefly in the recumbent posture. If the sore be slow in healing, it may be brushed over with a weak solution of the nitrate of silver.

The separation of the ligatures is very **rarely** indeed followed by bleeding. The only instance of trouble from this cause which has occurred to me, was the case of a robust, full-blooded gentleman, **aged fifty-six, for whom I** tied several large hæmorrhoids. He went on very well for a few days, and the last ligature came away on the ninth day. About four next morning as he lay in bed, bleeding took place into the rectum, and in a short **time he voided a** quart of blood and became faint. The local application of ice **arrested the bleeding, and there was no return of it, and he did quite well.**

The local irritation produced by the cautery or by the ligatures sometimes occasions retention of urine, and the passage of a **catheter may be required in the** evening after the operation. A hip-bath in addition to an opiate injection will generally relieve the urinary symptoms. They seldom last longer than forty-eight hours, though in one gentleman of irritable constitution upon whom I **had** operated, the catheter was required twice daily for a week afterwards.

In mentioning the operations which may be resorted to for the cure of internal hæmorrhoids, I have as yet made no allusion to the method by *écrasement linéaire*, introduced by M. Chassaignac, and I believe still preferred by many French surgeons. With the écraseur applied under anæsthesia a hæmorrhoidal growth can be removed in from eight to twelve minutes without pain, and without much risk of serious hæmorrhage, leaving a lacerated wound, which soon begins to heal. The chief advantages of this plan are the absence of pain after the operation, and the speedy cure of the patient. I have seen the operation performed by M. Chassaignac. Some little bleeding attended it, and I have read in the French journals accounts of three cases in which this operation has been succeeded by severe hæmorrhage coming on in the evening or next day. It seems, indeed, a rude operation, and when several hæmorrhoids require removal in succession, a tedious proceeding. The nice adjustment of the chain is also very difficult when the growth projects but little and has a broad base. As the operator inserts into the rectum a sort of multiplied harpoon to seize and drag down the growth, he is very apt also to include large portions of the mucous membrane around the piles. This leads to serious contraction of the anal aperture during cicatrization, constituting a very important objection to the operation. The late M. Nelaton stated in respect to *écrasement linéaire* badly done, " For a short time after its performance the patients are delighted, and the surgeon believes that he has attained a splendid result; but in the course of a few months the cicatricial tissue contracts, and the patient suffers from an anal stricture.

During about a twelvemonth I have seen a great number of patients, who have come to me in order to undergo an operation to remedy this unfortunate consequence of removal of hæmorrhoidal tumours, the stricture sometimes scarcely admitting the passage of a quill. It has arisen because not only the mucous projection which alone constitutes the disease has been removed, but also a more or less considerable portion of the skin of the orifice of the anus."[8] The above are important objections to the removal of hæmorrhoids with the ecraseur, and the operation is one which has found no favour in this country.

Ordinary bleeding from piles may be stopped by the introduction of pieces of ice into the rectum, or by injections of cold or iced water, or some astringent solution, as the sulphate of zinc or copper, or of the infusion of matico or rhatany; but when the hæmorrhage is continued, following every evacuation and weakening the patient, other measures must be taken to arrest it. I have already alluded to the prejudice which exists against interfering with bleeding piles, from the belief that the loss of blood is good for the general health, or that danger may be incurred in stopping an habitual discharge. On this ground bleeding is sometimes allowed to go on to an injurious extent before recourse is had to surgical assistance; but arterial hæmorrhage from piles is quite another matter from the occasional or periodical venous discharges to which many persons are liable. In cases of hæmorrhage, an effectual plan is to touch the bleeding point with strong nitric acid

---

[8] **Gazette** des Hôpitaux, 1860. No. 23.

in the mode already **described; or if** the pile should be large, **the surgeon** may seize **it or its bleeding extremity with the forceps or tenaculum, and include** the **part in a** ligature; **but, in consequence of an irritable** condition of the sphincter, there may **be** difficulty in exposing the bleeding point, and much opposition to the introduction of a speculum, without at least the assistance of chloroform. Under these circumstances, the surgeon may introduce a pencil of nitrate of **silver, and make** a free application of it to the surface, **which will** often **have the desired** effect, **or he may** apply a saturated solution of perchloride of iron **in glycerine. An enema of** cold water or an astringent injection, as the infusion of rhatany **root,** may be given just previous to the bowels acting. In hæmorrhage of an obstinate character, in which there is difficulty in discovering the exact source of **the** bleeding, **a** powerful styptic, consisting of two **drachms of tannic acid,** three of rectified spirit, and **seven of water, which I have used with** effect in **persistent epistaxis, may be injected into the rectum. The most distressing cases of piles met with in** practice **are those in which there is not only im**portant hæmorrhage, but ulceration on the surface, forming a painful and irritable sore. The following example will illustrate the difficulties of such cases, and indicate the treatment necessary in dealing with them.

Obstinate bleeding from an internal pile. Failure of caustics and astringents. Cure by operation.

Case 10.—On the 8th of November, 1845, a married lady, **naturally** of delicate constitution, and in impaired health from repeated miscarriages, noticed a rather free discharge of blood after a costive motion. She had been troubled by an internal pile for some years, and had just returned from the sea-side, where she had suffered severely from it. A dull pain had been felt in the lower part of the back after walking exercise. The uneasiness sometimes came on at night, lasting several hours, **and** disturbing her rest; she also suffered acutely for about an hour after every evacuation, and the **motion was** followed **by a pale yellow** discharge. Her spirits **became much depressed; she lost appetite, and returned to town in worse health than** when she **left home. In a few** days afterwards the bleeding occurred. **I made an examination**; and with some trouble, owing to the tender condition of the parts, got sight of a florid-looking pile rather deeply seated. I applied the lunar caustic pretty freely to the inflamed and ulcerated surface. The pain of the application lasted several hours. To my disappointment, there was a return of the bleeding after **the** next motion. Cold water and astringent injections were administered twice a day; but **the** bleeding still **continued.** A surgeon who saw the case **with me on** the 11th, suggested another free application of **the** caustic, which was made with some difficulty, owing to the painful spasm of the sphincter, but with no better effect than before. She took the acetate of lead, gallic acid, and had strong astringent injections of various kinds, but without arresting the bleeding, which, though not copious in amount, but occurring once or twice in the **twenty-four** hours for several days, **had rendered her anæmic, and much** reduced **the strength in her enfeebled state of health.** At a consultation with the late Sir B. Brodie, held on the 14th, it was determined **that an attempt should be made to tie the** pile. **The** resistance **of the irritable sphincter having** been overcome by force, and the lower part **of the** rectum partly everted, the pile was seized and dragged down, and with some difficulty a ligature was tightly secured round its base. **The pain** of the operation was excruciating, and it was necessary to give a drachm of laudanum immediately afterwards. The treatment, however, was quite effectual; there was no return **of the** bleeding, and the lady regained her usual health. Had the properties of anæsthetics been known at that time the treatment of the case would have been much facilitated, and the acute suffering altogether prevented.

Considerable difference of opinion exists as to which operation—the cautery or the ligature—is the best on the score of freedom from suffering, safety, and speedy recovery without confinement. As respects after-pain, there is little to choose, if care be taken to incise freely before the ligature is applied, and both operations are equally free from danger. The advocates of the cautery have, it is true, magnified the risks of the ligature. But, after a lengthened experience, I can state that, with one exception, no fatal case of operation by ligature has occurred either in my public or private practice. In 1868 a man, aged fifty-two, upon whom I had operated in the London Hospital for internal piles, and afterwards for fistula, was seized with pyæmia, followed by numerous abscesses, and, after a protracted illness of five months, died. Erysipelas and pyæmia have also followed, though very rarely, the operation by cautery. Some amount of danger must be incurred in every kind of operation, serious results sometimes arising from the slightest causes; and the removal of piles cannot be expected to be exempt from risks which may attend a trifling puncture of the finger.[9] An unfortunate result after

[9] In the spring of 1858 no less than four persons died of tetanus in St. Mark's Hospital after operations for hæmorrhoids, but at that particular period attacks of tetanus after operations and accidents were unusually prevalent in London, and no case of the kind has occurred since in the special hospital alluded to. In 1873 I saw in consultation a publican, aged thirty, who was seized with

an operation for piles, in persons free from organic disease, is, however, entirely exceptional, and, with common precautions, it may be regarded as safe as any operation in surgery.

I continue to give the preference to the ligature as more convenient to the surgeon and less alarming to the patient. The operation by cautery is more tedious than by ligature, a matter of little moment when the patient is insensible, and special care is necessary in using the cautery to guard against after-hæmorrhage. In a favourable case the patient is able to leave the couch somewhat sooner, and recovery is, on the whole, somewhat more rapid than after ligature, as the healing of the sores does not commence till after the separation of the ligature, a process occupying four or five days, whereas the eschar, or superficial slough produced by the cautery, is cast off sooner. A trouble liable to occur after the removal of piles arises from the sores assuming the characters of an irritable ulcer, which I met with in two instances some years ago. As a rule, I avoid operations for internal piles after the age of seventy. I attended with Dr. Wynn Williams an old lady, aged seventy-three, with cataract, who was also troubled with bleeding piles. The hæmorrhage rendered an operation on the eye unpromising. I removed the piles, and she recovered slowly, but favourably. Mr. Critchett subsequently extracted the opaque lens with a good result.

The symptoms of hæmorrhoids appear to admit of some temporary relief from mechanical pressure.

acute tetanus after the removal of some large piles by ligature by a skilful surgeon of great experience. The case terminated fatally. I have also heard of another fatal case of tetanus after ligature of piles in the Metropolitan Free Hospital.

For this purpose, bougies have been introduced into the rectum, and retained there for a certain period daily; and short metallic plugs have been worn for the same purpose. The principle of giving support to weak and dilated veins by mechanical means is of the utmost value in surgery, but it is obviously impossible to apply this with effect to the hæmorrhoidal veins; and, though some benefit may be derived from the instruments alluded to, the relief is too slight and transient, and the treatment of too disagreeable a character to render these plans of much value in practice. In persons, however, advanced in life, with a weak sphincter and relaxed rectum, especially in men with enlargement of the prostate gland, the hæmorrhoidal growths and the adjoining mucous folds slip down so readily, when the patient stands or walks about, that it is often necessary to adopt some mechanical means to support them, and prevent their protruding. Such measures are chiefly required in cases where the patient is very reluctant to undergo an operation, or is an unfit subject for one. There is an instrument in common use, consisting of a steel band to encircle the pelvis, from the back of which a slightly curved spring descends to a point corresponding to the anus, at which extremity a conical ivory or india-rubber pad is attached. The pad, pressed upwards by the spring, supports the rectum, into which it slightly enters. Some of these instruments possess the advantage of not being exposed to displacement in the changing positions of the body, the back spring being only loosely attached to the circular band. I can conceive, however, nothing more disagreeable than constantly to feel at the orifice of the anus a foreign body which is always attempting to penetrate the rectum. The remedy is almost worse than the disease. Perhaps a more comfortable support is that given by an air or water pad connected to an abdominal bandage in front, and behind by elastic bands. A plug with a contracted neck, the part grasped by the sphincter, sometimes answers the purpose of supporting the bowel, and can generally be worn with comfort.

CHAPTER VI.

PROLAPSUS OF THE RECTUM.

In describing the changes occurring in piles, I remarked that internal hæmorrhoids slip down and project at the anus. The descent of these growths is often attended with more or less eversion of the hypertrophied mucous membrane of the lower part of the rectum, similar to what takes place, although in a slighter degree and only temporarily, in the ordinary actions of defecation. In relaxed states of the sphincter muscle and coats of the bowel, loose folds of mucous membrane alone are liable to protrude and to require replacement. This protrusion and exposure of the thickened mucous membrane, with or without internal hæmorrhoids, have been erroneously described by writers as prolapsus of the rectum. In the true prolapsus, however, there is a great deal more than an eversion of the lining membrane of the bowel. The gut is inverted; there is a "falling down" and protrusion of the whole of the coats—a change in many respects analogous to intussusception, but differing from it in the circumstance that the involved intestine, instead of being sheathed or invaginated, is uncovered and projects externally.

The length of bowel protruded in prolapsus varies greatly, from an inch to six inches, or even more. The shape and appearance of the swelling depend partly upon its size, and partly upon the condition of the external sphincter. When not of any great length, the protrusion forms a rounded swelling, which overlaps the anus, at which part it is contracted into a sort of neck. In the centre of the swelling there is a circular opening, communicating with the intestinal canal. An inversion of greater extent usually forms an elongated pyriform tumour, the free extremity of which is often tilted forwards or to one side, and the intestinal aperture assumes the form of a fissure, receding from the surface of the tumour, owing to the traction exerted upon it by the meso-rectum. In a relaxed condition of the sphincter the surface of the protrusion has the usual florid appearance of the mucous membrane; but in other cases it is of a violet or livid colour, and tumid from congestion, the return of blood being impeded by the contracted sphincter. The exposed mucous membrane is often thickened and granular, and sometimes ulcerated from friction against the thighs and clothes. A thin film of lymph may be occasionally observed coating its surface. On examining a section of a large prolapsed rectum taken from the body of a child, I found the coats of the protruded bowel greatly enlarged; the areolar tissue was infiltrated with an albuminous deposit, the muscular tunic hypertrophied, and the mucous membrane much thickened and dense in structure, especially at the free extremity of the protrusion. These changes account for the difficulty in reducing the parts, and in retaining them afterwards, so often experienced in the treatment of these cases in children, the bowel having become too large to be conveniently lodged in its natural position, and, like a foreign body, exciting the actions of expulsion. I believe that in all cases of prolapsus, in which the parts are suffered to remain unreduced for any time, the coats of the extruded bowel will be found thickened and hypertrophied, from the irritation to which they are exposed in this condition. The atonic and relaxed state of the sphincter muscle is well shown by the facility with which one or two fingers can be passed through the anus even in young children.

Prolapsus of the rectum is observed most frequently in children between the ages of two and four, but is liable to occur at a later period in life. In infancy it is produced by protracted diarrhœa; the frequent forcing at stool so weakening the coats and connexions of the rectum, and relaxing the sphincter, as at length to lead to inversion of the bowel. These young subjects are generally in feeble health. The straining efforts to pass water, consequent upon stone in the bladder, also often give rise to this affection in early life. In adults prolapsus is said to be produced by too copious enemata, but the descent results chiefly from a weakened condition of the sphincter and levator ani muscles, and a general relaxation of the tissues of the part. The rectum being imperfectly supported by the perineum, the eversion at stool gradually extends, until an actual inversion takes place, which may increase until it forms a protrusion of considerable size. In adults prolapsus is more common in women than in men. In the former, it results in a great measure from weakness in these parts produced by repeated child-bearing. The extent to which the sphincter sometimes admits of dilatation in women, and the amount and size of the parts falling through it, are really remarkable. There is a preparation in the Museum of the College of Surgeons,[1] consisting of a considerable portion of the rectum inverted and protruded through the anus, forming a tumour of nearly hemispherical form, between three and four inches in diameter. The mucous membrane of the rectum is thickened and extensively ulcerated; the opening through which the parts are protruded is of great size; there is also some degree of prolapsus of the uterus, with inversion of the vagina. Some years ago I was asked to visit a poor Jewess in Petticoat-lane, the mother of several children, who had a prolapsus of the rectum which formed a round tumour the size of a child's head of two or three years of age. There were large ulcers on the surface; the anal orifice was dilated to an enormous extent. She was in a miserable condition, being unable to pass her water or to evacuate her bowels without forcing up the protruded parts, which slipped down the moment her hands were removed from the swelling. A prolapsus may be combined with internal piles. We meet with this in men affected with enlargement of the prostate or stricture, and who are accustomed to strain in passing water.

[1] No. 1382.

This frequent forcing, as well as the habitual protrusion of the hæmorrhoidal folds, so weaken the sphincter and relax the coats and connexions of the rectum, as ultimately to cause displacement and inversion of the bowel. In these cases the hæmorrhoids will be observed encircling the upper part of the protrusion near the anus.

The annoyance and inconvenience occasioned by a prolapsus of the rectum vary considerably under different circumstances. Thus the bowel may descend only in a very slight degree at stool, and disappear by a natural effort afterwards; or it may come down only occasionally, admitting of being easily thrust back, and, when returned, will remain in its place until an attack of diarrhœa, or the effort to pass a costive motion, causes it to fall again. Prolapsus sometimes occurs after every motion, and even when the patient stands and moves about, forming a large red unsightly tumour exposed to friction, feeling sore, and soiling the linen with a bloody discharge, and requiring to be pushed up frequently during the day. Or the gut may be constantly protruded, being fixed so as not to admit of replacement. There are cases on record in which a prolapsed bowel has become strangulated and inflamed, and has even mortified and sloughed off; similar to what sometimes happens to an invaginated intestine.

Large prolapsus of the rectum, with sloughing of the mucous membrane.

Case 11.—An emaciated woman with an anxious expression of countenance, aged 61, was admitted into the London Hospital in November, 1861, under my care, on account of a large prolapsus of the rectum. She had been subject to it for some years, and it first occurred after lifting a heavy weight. It came down two days before her admission, and not having succeeded in returning it, she became alarmed. I found the protruding bowel swollen and gorged, forming a large pyriform swelling the size of an orange, of a dark colour, its surface being covered to a great extent with a black slough. The margin of the anus tightly constricted the base of the tumour. The patient suffered a good deal of pain in the part. I applied a sponge over the protrusion, and by firm and continued pressure endeavoured to reduce it, but did not succeed in making any impression on the tumour. She was kept in bed, and large superficial sloughs separated from the mucous surface of the exposed bowel in about a fortnight, when, the engorgement having subsided, I was able to return the protrusion by gentle pressure without difficulty. On digital examination after a week, I detected a slight contraction in the passage consequent on the healing of the sores, but this yielded readily to dilatation with bougies, and the patient was discharged completely cured seven weeks after her admission.

All the worst forms of prolapsus in early life that I have met with have been amongst the neglected children of the poor. Young persons generally outgrow this complaint by the period of puberty; and, common as prolapsus is in early life, it is rather rare in young grown-up subjects. I have known, however, of persons who have had this disease in infancy, becoming affected with a return of it in later life from the effects of a diarrhœa.

In adults prolapsus is commonly attended with a slimy discharge of mucus tinged with blood, and, in some instances, with troublesome bleeding. The hæmorrhage does not occur from any particular spot, but as an exudation from the congested mucous surface when the bowel is protruded at stool. As the cause producing the hæmorrhage is constantly recurring, there is sometimes considerable difficulty in arresting it, local applications having little effect so long as the bowel continues to descend. In an obstinate case of this kind in a gentleman about forty years of age, who was quite pallid from the continuance of small losses of blood in this way, I applied the nitrate of silver and sulphate of copper to the exposed mucous surface, and used various astringent injections before the bowels acted, but without any effect in stopping the bleeding until I succeeded by treatment in preventing the bowel falling. I have experienced similar difficulty in arresting the bleeding in other cases so long as the prolapsus continued.

In children, irritability of the bowels and diarrhœa must be checked, and disordered secretions corrected by suitable remedies. Attention must be paid to diet; and when the powers are feeble, benefit will be derived from quinine or steel medicines. Cod-liver oil often proves of great service in causing easy motions, and restoring the general health at the same time. In slight cases it will be sufficient to direct the nurse, when the rectum descends at stool, to place the child on its face across her lap, and to return the parts by taking a soft cambric handkerchief or sponge, wetted in cold water, in both hands, and by gentle, but steady compression, to push the protrusion back into the pelvis. The relaxed state of the membrane may be corrected by administering regularly every evening an astringent injection, which may remain in the bowel during the night. I usually prescribe the decoction of oak bark, with alum, in the proportion of a scruple of the latter to eight ounces of the decoction, a third of the quantity being sufficient for use. The infusion of rhatany is often used with advantage. From twenty to thirty minims of the muriated tincture of iron, added to four ounces of water, also makes an excellent astringent enema for these cases. The injections should be used cold. If the bowel should slip down when the patient moves about, mechanical support must be given to the part. A well-fitting rectal supporter, worn constantly for a certain period, will be of great service in maintaining the bowel in its place. When from swelling and thickening of the coats of the rectum the intestine becomes almost fixed in its unnatural position, there is greater difficulty in the management of the case. Continued and pretty strong pressure will be required to replace the bowel. If the struggles of the child should cause much resistance to the efforts of the surgeon, chloroform inhalation will facilitate matters, and have a good effect in relaxing the sphincter, rendering unnecessary the division of the muscle, which has been recommended in cases, both of children and of adults, where much difficulty is experienced in replacing the bowel. When the exposed surface is ulcerated, benefit may be derived from painting the diseased part with a solution of the nitrate of silver. The chief difficulty, however, is to retain the parts after they have been reduced. A good-sized piece of sponge, or a piece of cotton wool moistened in a strong solution of extract of rhatany (gr. x.—ʒj.), may be lodged at the anus, and firmly secured there by approximating the buttocks, by means of a broad strip of adhesive plaster applied across from one side to the other, and further secured with a T bandage. This will require to be readjusted after every motion. The child should also be kept at rest in bed, and be made to relieve its bowels in the recumbent posture, until the strong tendency to prolapsus has been corrected by this treatment, coupled with astringent injections. Afterwards the usual bandage may be applied, and the patient allowed to move about. In cases of children with stone in the bladder, the prolapsus generally disappears spontaneously after the operation of lithotomy, and the removal of the original cause of the complaint.

In cases of prolapsus in adults, accompanied with external or internal hæmorrhoids, the contraction that takes place after the removal of the piles by excision, ligature, or in other ways, will often counteract the laxity of the parts, and afford sufficient support to prevent a return of the inversion. The efficacy of this treatment was first made known by Mr. Hey, who published, in his "Practical Observations in Surgery," an interesting series of four cases in which a prolapsus attended with bleeding was cured by excision of hæmorrhoidal excrescences. In the instance of the woman in the London Hospital, whose case is related at page 70, the process of healing after the sloughing of the mucous membrane and submucous tissue on the surface of the protruded bowel, caused so much contraction as not only effectually to cure the prolapsus, but even to produce a slight stricture in the bowel.

The contraction necessary to prevent the fall of the rectum may also be obtained by the application of escharotics, such as the mineral acids or the potassa fusa, to the mucous membrane near its junction with the skin, so as to form superficial sloughs of greater or less extent according to the amount of laxity to be counteracted. In slight cases of prolapsus this treatment is very effectual and not very painful. Guersant, a French surgeon, had recourse to the actual cautery for the formation of the sloughs required, and recommended this treatment for prolapsus in children. It appears that the cautery often caused painful sores, which healed with difficulty. The application of escharotics has been recommended to the surface of the prolapsed bowel, instead of to the mucous membrane only near the anus. This must be done with great caution, for if the coats of the bowels are destroyed to any depth serious contraction is very liable to ensue. A young gentleman with prolapsus treated with nitric acid in this way was brought to me on account of firm annular stricture about three inches from the anus, through which I could pass only the tip of my finger, the prolapsus still continuing. By the persevering use of bougies during eighteen months the stricture was dilated and completely cured, and then the prolapsus disappeared. In severe cases of prolapsus the fall of the rectum may be effectually obviated by an operation which consists in the excision of portions of mucous membrane, and of skin from the margin of the anus. The patient being placed on his back in the position usual in the operation for lithotomy, a fold of membrane, more or less broad according to the laxity of the part, is to be seized with a volsellum, or the hæmorrhoidal forceps, raised a little, and then excised with a curved pair of scissors. Two portions, one from each side of the rectum, will generally require removal, leaving two oval wounds in the longitudinal direction. It is desirable that the edges of the wound should be afterwards brought together with sutures, not only to secure the speedy healing of the wound, but to prevent bleeding by compression of the parts. Unless, however, anæsthesia be produced, there is some difficulty in applying them, in consequence of the forcible contraction of the sphincter excited by the operation drawing in and concealing the wounded parts. The surgeon must be careful to tie any bleeding vessel that may be divided, for the operation is very liable to be followed by hæmorrhage, which may go on into the bowel without his being aware of it. Cases in which this operation is called for are not very common. In persons who have suffered from prolapsus in childhood it sometimes happens that the parts do not recover their tone at puberty, and that the complaint continues to prove troublesome afterwards. Such a case is very fit for cure by excision. Many years ago, before the use of anæsthetics, I assisted Mr. Luke in performing this operation upon a lad in the London Hospital. He was nineteen years of age, and had been troubled with prolapsus ever since he was three years old. The bowel always descended several inches when he went to stool, and was a source of great annoyance to him. Two oval portions of mucous membrane were excised from the verge of the anus in the way above described; but the wounds were not closed with sutures. The sphincter immediately afterwards contracted strongly, and completely buried the wounded surfaces. There was no reason at the time of the operation to expect any bleeding; but on visiting the lad in the evening I was surprised to find him in a state of prostration, with a cold, clammy skin, and shivering. It appeared that on two or three occasions he had discharged a considerable quantity of blood, which had collected within the rectum. Having given him some brandy, I introduced a thick plug of lint, previously oiled, which was effected with some difficulty, owing to strong spasm of the sphincter. There was no recurrence of hæmorrhage, and the two wounds healed up in the course of a month. The operation was quite successful in preventing further prolapse. Had the wound been closed with sutures, most probably no bleeding would have occurred afterwards.

The operation of excision is also applicable for the cure of prolapsus in women from a weakness of the parts consequent on child-bearing. This weakness is sometimes so great that the fæces, when fluid, escape involuntarily. In these cases, as there is considerable dilatation or elongation of the sphincter, it has been proposed to shorten the muscle, by excising a portion of it on each side. The operation is not difficult. The anal ring must be grasped with a sharp hook or volsellum, and a wedge-shaped portion excised with a small scalpel. The wounds must afterwards be closed by sutures.

In many instances, the advanced age or state of the general health of the patient renders an operation of any kind inadvisable. A proper rectum supporter will help to lessen the inconvenience; and should difficulty be experienced in returning the protrusion, and the patient be obliged to lie down in order to effect it, comfort will be derived from his establishing the habit of relieving his bowels the last thing at night, so that he may retire to rest at once, and remain in a position favourable for the reduction, and for the prevention of the prolapsus until the morning.

CHAPTER VII.

POLYPUS OF THE RECTUM.

When considering the changes consequent upon hæmorrhoids, I described the hypertrophied folds, developed in this disease, as sometimes assuming an elongated form, and protruding at the anus. These processes rarely become pedunculated, but spring from the lower part of the rectum, just within the external sphincter, and are usually attached by a broad base. Growths, however, occasionally arise from the mucous membrane of the rectum higher up in the passage, being attached by a narrow and elongated pedicle. A tumour of this kind is called a *polypus of the rectum*.

Polypus of the rectum occurs in two forms, the *soft* or *follicular*, and the *hard* or *fibrous*. The soft polypus forms generally in early life. Its essential element is a considerable agglomeration of elongated follicles containing, according to Lebert,[2] a very distinct cylindrical epithelium, the epithelium of the intestine at this part being pavement. There is a network of small vessels on its surface, which is also furnished with beautiful papillæ. The peduncle varies in length. The polypus is usually single, but several

---

[2] Traité d'Anat. Pathologique, vol. i p. 268.

may form. Lebert met, in one instance, with as many as twenty in a girl aged seventeen, who was under treatment by M. Robert in the Hôpital Beaujon.

In children the follicular polypus usually makes its appearance external to the anus after a stool, resembling a small strawberry, being of a soft texture, granular on its surface, and of a red colour. It has a narrow pedicle about the size of a crow's quill, and two or three inches in length, by which it is attached to the interior of the rectum, usually the posterior wall. It produces scarcely any suffering, but causes a slight bloody discharge, which, appearing after every motion, excites some alarm. In some instances the bleeding is sufficiently free to weaken the child.[3] The description of the complaint given by the mother or nurse is liable to mislead the practitioner, and to induce him to conclude that the case is the more common affection—prolapsus. The real nature of the complaint can generally be detected by the introduction of the finger into the rectum. But it sometimes happens, owing to the polypus being very moveable, that it passes up the bowel and gets out of reach. When the peduncle is long enough, the growth is forced out at stool, and its nature can then be ascertained without difficulty. The follicular polypus occurs very rarely in the adult.

The treatment of polypus in children is very simple, and always effectual. The tumour should

---

[3] Dr. Woodman, who appears to have seen many of these cases at the North-Eastern Hospital for Children, describes the patients as being troubled with a bearing down and crying at stool, and also as suffering from a teasing diarrhœa.—*Medical Press and Circular*, Nov. 24, 1875.

be strangulated by a ligature secured around the pedicle, and then returned within the bowel. This gives no pain, and produces no suffering afterwards, and the polypus separates and comes away with the motions in the course of two or three days. If the peduncle be long, the polypus may be cut off beyond the ligature.

Case 12.—A child four years of age had a soft polypus the size of a nut. It had a long peduncle, and projected externally after motions. I placed a tight ligature round the root of the peduncle, and then excised the tumour. The child ran about as usual, and the part got well as if nothing had happened.

A polypus should not be excised without the previous application of a ligature, as dangerous bleeding is liable to occur from the cut surface of the pedicle. This happened, in a case operated on by Sir A. Cooper, to such an extent as to occasion alarm. Nor should the ligature be tied so tight as to divide the soft neck, for hæmorrhage has been known to arise from this cause. Mr. Mayo mentions, that in tying a polypus of the rectum in a girl eleven years of age, he drew the ligature so tightly that it cut through the slender pedicle. There was no bleeding at the time, but the following night the child lost a profuse quantity of blood, and came to the hospital the following day faint and pale, and reduced from the bleeding.[4]

Case 13.—A boy, about five years of age, was brought to me with a soft polypus in the rectum. The growth being high up the bowel, I found it impossible to get a noose

[4] Outlines of Human Pathology, p. 354.

round it. The child was under the influence of chloroform, but in so young a person I could introduce only one finger into the rectum to manipulate with, and was unable to drag the polypus out of the gut. It got completely broken down and destroyed under the attempts made to tie it. There was, however, no bleeding of any account at the time or afterwards, and the growth did not return.

The following case will serve to illustrate some of the chief points of practical interest in these cases.

Case 14.—A little girl, of sickly appearance, was brought to me in consequence of a swelling protruding at the anus after stool. The nurse described it as resembling a cherry, and stated that it constantly presented after an evacuation, and often required to be pushed back into the passage. It caused no uneasiness, but was attended with a slight bloody discharge. I was unable to induce my little patient to make any straining effort to cause the body to project, and on introducing my finger into the rectum could feel no swelling of any kind. As the parents resided twelve miles out of town, there was difficulty in getting an opportunity of examining the part after a stool. Apprehending that the case might be prolapsus, I prescribed steel medicines, and directed the tumour to be returned with a piece of soft lint, wetted with a solution of sulphate of zinc. I subsequently ordered an injection of the muriated tincture of iron to be administered daily. After paying me two or three visits, the child was taken to the sea-side for the improvement of its general health, and brought to me again on her return. Finding that the projection and discharge were not diminished, I made another examination with the finger, but could find no tumour. I ordered a dose of castor oil to be given early in the morning, and the child to be brought to me afterwards. After she had remained in my house an hour or two, the bowels acted, and I then succeeded in getting sight of a dark red vascular tumour, the size of a small cherry, which protruded at the anus, and had a long narrow pedicle. I passed a ligature round this without difficulty, and returned the strangulated swelling into the rectum. No suffering was produced; and in three days the tumour came away at stool, and the child was cured.

The hard or fibrous polypus occurs in adults. It originates probably in hypertrophy of the submucous areolar tissue of the rectum. Its surface is in some cases smooth, and in others irregular. I examined a pedunculated tumour of this kind removed from a man by operation. It was of an oval shape, and the size of a chestnut, had a firm stem about the diameter of a goose's quill, and the irregular nodular surface of a cauliflower excrescence. It was composed principally of fibrous tissue.

The fibrous polypus is of a pear shape, with a peduncle more or less long and thick. It varies in firmness, seldom bleeds, but occasions a slight mucous discharge, and when the peduncle is long or the tumour low down, it protrudes at the anus after stool, and requires replacement. When lodged within the bowel it causes a sensation of unrelief, as if a foreign body or feculent lump required discharge. The polypoid growth sometimes becomes congested, and when protruded in this state its peduncle is liable to become girt by the sphincter, which causes great pain. I attended with Mr. Langmore a highly nervous lady who had a growth of this description attached about an inch and a half above the sphincter. It was subject to enlargement under congestion, and on replacement in this condition severe pain came on an hour or two afterwards. She was relieved by an operation.

Case 15—Many years ago I was requested to visit in the country an esteemed member of our profession who was in great suffering. He had been troubled with a fibrous polypus for some years. Probably from being chafed whilst protruded it became ulcerated on the surface. This set up distressing spasm of the sphincter, which constricting the peduncle produced congestion of the tumour. His tortures were extreme. I found a large dense pear-shaped body, with a thick and firm peduncle, attached only a short distance within the bowel. There was a large ulcer on the mucous surface, which was of a dark red colour. I transfixed the peduncle with a needle armed with a double ligature, which was tied tightly on each side. The tumour sloughed off, and the patient was speedily and effectually relieved of all pain, and was quite cured of a most annoying complaint.

If the peduncle of a fibrous peduncle be small, it may be encircled by a tight ligature, and be left to slough off, unless there is room for excision beyond the ligature. When the peduncle is large and strong, the wire écraseur answers best.

Case 16.—An elderly lady was brought to my house by her physician on account of a considerable polypoid growth in the rectum. With some trouble the growth was dragged outside the anus, and having no écraseur at hand, I applied a strong twine ligature as tightly as possible round a large and firm peduncle. No suffering ensued, and a fortnight afterwards I discovered that the ligature had rotted away, having caused a reduction of about a third of the growth, which was sloughy on the surface. I then applied the wire écraseur and effectually removed the polypus. She was going on well, when on the fourth day after the operation she was seized with rigours. Symptoms of pyæmia ensued, and she died ten days later. This is the only case of pyæmia after an operation connected with the rectum which has occurred in my private practice after a long experience.

Owing to the fibrous polypus being attached high up in the rectum, or to its peduncle being short, it may be difficult or impossible to draw down the growth so as to get a ligature or the écraseur around the neck. In that case a clamp forceps slightly curved may be applied to the peduncle within the rectum, and firmly secured by a screw in the handle, so as to tightly compress the part and strangulate the growth. The bowels should afterwards be kept confined with opium until the forceps comes away.

A firm polypus attached low down in the rectum sometimes originates in an elongated pile, the extremity of which has undergone conversion into fibrous tissue. This fibrous growth occasionally sets up irritation in the mucous membrane, and causes an ulcer within the circle of the sphincter. The sore assumes the characters of the irritable ulcer, and the suffering is severe, the polypus coming in contact with the ulcer and disturbing it. The polypus should be removed and a slight incision made across the ulcer to facilitate its cure.

CHAPTER VIII.

VILLOUS TUMOUR OF THE RECTUM.

A GROWTH similar to the villous tumour which occurs in the bladder and on other mucous surfaces, sometimes forms in the rectum. It springs from the mucous membrane generally by a broad base, is soft in structure, and composed of a number of projecting papillæ or villi. On minute examination it is found to vary in structure according to the proportion of the fibrous or vascular elements entering into its composition. A characteristic specimen in the London Hospital Museum is described by Dr. A. Clark as essentially an outgrowth of dense areolar tissue, permeated by blood-vessels, and assuming a papillary form, the papillæ being flattened and curled so as to represent hollow cylinders, and being clothed with layers of epithelium, the free layers being cylindrical. Its minute structure indeed closely resembles the soft polypus. The villous tumour is innocent in character, and is not apt to return after complete removal. Its chief peculiarity in the rectum, as in the bladder, is a remarkable disposition to bleed. This growth occurs only in adults, and is a rare disease. When it projects at the anus, it exhibits the characteristic projecting processes of a deep red colour.

Mr. Syme has described two cases of soft vascular bleeding polypus of the rectum which, I presume, was this disease. He states, that he removed from a hospital patient a tumour not less than an orange, which had nearly exhausted the patient by hæmorrhage. In another case, in which the disease was detected from the great hæmorrhage which it occasioned, he tied the tumour within the rectum.[5] The villous tumour has not generally a peduncle or neck. It has been particularly described by Mr. Quain under the name of a "Peculiar Bleeding Tumour of the Rectum," but as it closely resembles the outgrowths found in the bladder, usually called *villous*,

---

[5] Diseases of the Rectum, 2nd edit., p 82.

I prefer the latter term. Mr. Quain met with it in two cases in females, one middle-aged, the other sixty-eight.

The largest tumour of the kind which I have seen **was in a man of** middle age, a patient at St. Mark's Hospital, and was shown me by Mr. Gowlland. The growth was attached to the **lower part of the rectum** by a broad base, and projected at stool. It had been forming for some **years.** This tumour was successfully excised by operation, and it is the one already referred to as examined by Dr. A. Clark.

The bleeding to which the villous growth generally gives rise, and the slimy discharge, render its removal very necessary. If the tumour be attached high up, and a ligature can be applied around the base of it, this is desirable, as it would be difficult to arrest the bleeding after excision. If the growth cannot be strangulated by ligature, the clamp forceps may be applied to its base.

CHAPTER IX.

FISTULA IN ANO.

The loose areolar tissue around the lower part of the rectum is occasionally the seat of abscess, which bursts externally near the anus. But instead of the part healing afterwards, like abscesses in other situations, the **walls** contract and become fistulous, and the patient is annoyed by a discharge from **the opening.** Such is the complaint termed *fistula in ano;* and though a very common **disease,** and one, apparently, **of very simple character,** there are still **some points connected with it, respecting which a difference of opinion exists.**

The abscess giving rise to fistula sometimes forms with all the characters and symptoms of acute phlegmon, suppuration taking place early, and the matter coming quickly to the surface. But more frequently a thickening appears at a spot near the anus with scarcely any sign of inflammation, and but little local pain, and is gradually resolved into a **fluctuating swelling, which, being opened, discharges a fetid pus.** On introducing **a probe at** the external orifice of a fistula formed **in either** way, it may pass through a **small** opening in the coats of **the rectum into the bowel.** The case is then called a *complete fistula.* When there is no external **opening, the** complaint is named *blind external fistula.* The **external orifice is** usually but a short distance from **the anus,** its situation **being often indicated by a button-like growth; and it is in the centre of this** red projecting granulation **that the opening is found.** The aperture, however, **is not always so marked, and being very small, a mere slit concealed in the** folds of the anus, it cannot be detected without careful search. The course taken by a fistula varies a good deal. I have a preparation in which the opening is so close to the margin of the anus that **the** sinus traverses the substance **of the external sphincter,—a course which is not, indeed, very uncommon.** The abscess, before breaking or being opened, **may, however, have** burrowed to some distance, and the external orifice may be placed two or three inches off **in the** direction of the buttock or perineum.

Fistula in ano arises in different ways. It commonly originates in the areolar tissue, near the anus, in a phlegmonous abscess consequent on congestion and inflammation; the frequent action of the sphincter muscle, and the disturbance of the part in defecation, preventing the closure of the sac in the usual mode. This does not, however, always happen. Some years ago I was asked to examine a robust, middle-aged professional friend, who was troubled with an abscess which had recently burst near the anus. I introduced a probe, and found by the finger in the rectum that it passed close to the mucous membrane of the bowel. I stated that he would require the operation for fistula, but requested him to remain quiet, and to wait a week. On my next visit I found the abscess closed, and the part quite sound. I have since met with many similar cases, in which, by making a timely opening, by rest afterwards, and mild aperients to prevent hard evacuations, an abscess which threatened to terminate in fistula has quite healed. When a sinus forms, it burrows close to the outer surface of the mucous membrane of the rectum, which forms a thin barrier between the bowel and the sinus. This shortly ulcerates, and thus is formed the internal orifice of the fistula. But this does not invariably take place. I have, in a few instances, met with a fistulous opening near the anus in which no communication with the bowel could be found on the most careful examination. That such a fistula occasionally occurs I have no doubt, notwithstanding the opinion of so high an authority as the late Sir B. Brodie, who, in a valuable lecture on this subject,[6] states that he is satisfied that the inner opening

* The *Lancet,* 1843-4, vol. i. p. 592.

always exists. I have observed one fistula of the kind in the dead body; and a few preparations showing the same fact may be seen in our hospital museums.[7] The abscess may make its way into the bowel before bursting externally, but the inner opening is commonly formed subsequently to the outer, and is small in size. When a fistula originates in active inflammation in the way above described, there is a sensation of weight about the anus, swelling of the integuments, considerable tenderness on pressure, pain in defecation, sometimes retention of urine, and constitutional disturbance, with rigors. These symptoms are relieved after the matter is discharged. The congestion to which the hæmorrhoidal veins are very liable, is, I have no doubt, the principal cause of the abscesses in the vicinity of the anus, inflammation and its consequences being readily produced in parts so favourably formed for such disease.

A sore formed in the little pouch, just within the external sphincter, and originating in the irritation to which this part is liable, instead of spreading superficially, sometimes perforates the bowel, and allows the escape of a little feculent matter into the areolar tissue around it. I attended with the late Dr. Ashwell a young married lady, who had an affection of the rectum. On examination with the speculum, we detected an ulcer of the mucous membrane at the lower and back part of the rectum. A fortnight afterwards an abscess pointed near the anus, and ended in a complete fistula, which opened internally at the seat of the ulcer. A very similar case is related by Sir B.

[7] Vide preparations numbered 35 and 46, series xvi., in the collection at St. Bartholomew's Hospital.

Brodie. Some years ago I operated for fistula on a patient of Mr. Arthur, of Shadwell, a married woman, who had suffered more than usual from the complaint. The wound healed in a fortnight; and on examining the part carefully, in consequence of her still suffering considerable pain, especially after defecation, I detected an ulcer at the back of the rectum, a short distance only from the inner opening of the fistula. Mr. Arthur attempted to cure this by different applications, but without success; and at the end of a month I divided the ulcer and sphincter muscle, after which the sore healed. In this case it appears that two separate ulcers formed in the rectum. One perforated the bowel; the other remained a painful superficial sore. Again: ulceration induced by an internal pile, and more rarely by a pointed foreign body, as a fish-bone, sticking in the mucous membrane, may produce perforation, and a rectal abscess. I once operated on a fistula originating in the impaction of a fish-bone, which had produced very extensive suppuration in the buttock and perineum. In all these cases the inner opening is found just within the external sphincter; indeed, in whatever way a fistula originates, this is the most usual situation for the orifice. This point was established many years ago by M. Ribes,[s] who examined a large number of bodies in order to ascertain the precise situation of the inner opening. In seventy-five subjects he never found the opening seated higher in the rectum than five or six lines: in a certain number it was at a distance of only three or four lines. M. Ribes' observations clearly show that the inner opening of the fistula is, in a large majority of

[s] Quarterly Journal of Foreign Medicine and Surgery, vol. ii., 1820.

cases, a very short distance only from the margin of the anus, and they are fully confirmed by Sir B. Brodie, who, indeed, goes so far as **to say,** " **The** inner orifice is, I believe, always situated immediately **above the** sphincter muscle, just the part where the fæces are liable to be stopped, and where an ulcer is most likely to extend through both the tunics." This, however, I have by no means found to be so constantly the case. I have examined several patients with fistula, and inspected the parts in others after death, in which the opening into the bowel **was** more than an inch above **the** external sphincter. There are several preparations of the kind in the London Museums.

Fistula occurs in phthisical subjects, originating in tubercular ulceration of the mucous membrane, and perforation of the bowel. In these cases the inner orifice is usually large in size, and there is **sometimes a second opening.** It is somewhat remarkable that Andral and Louis should have found this complaint very **rarely indeed in phthisis, when** all surgeons agree that fistula **is by no means of unfrequent** occurrence in **patients afflicted** with **tubercular** disease of the lungs. The abscesses originating in ulceration of the mucous membrane often form insidiously, patients suffering but little constitutional disturbance, and scarcely any local uneasiness, until the abscess is near the surface, and about to **burst. In other** instances the **symptoms are severe: there are** rigors, and considerable **febrile derangement, sometimes of the** low type usually **attending** the formation of fetid abscesses.

Though the inner orifice is very commonly found just within the external sphincter, communicating with one of the little sacs situated at this part, the fistula itself often extends some distance up the side of the rectum, as far as two or three inches, or even higher; and it may burrow in different directions. Formerly, surgeons, in examining patients, not being able, on passing the probe up these sinuses, to find any opening into the rectum, used erroneously to conclude that there was no communication with the bowel,—that the fistula was a blind one: but since the anatomy of the disease has been better understood, and greater pains have been taken in the examinations, search being made in the right direction, an inner opening has generally been detected. When the sinuses are tortuous or pass in different directions, there is sometimes more than one inner opening. There may be one in the usual situation, and another higher up, or on both sides of the rectum with an indirect communication between the sinuses. Sometimes there is an external orifice on each side of the anus leading to fistulous passages, which pass to the back of the rectum, and communicate with the gut at this part by a single orifice, so as to form a sort of *horse-shoe fistula*. The matter is liable to lodge in these complicated sinuses, to give rise to inflammation, and to lead to fresh abscesses and additional fistulous passages. When the disease is of old standing, the sides of the fistulous passages are often dense and callous, feeling gristly to the finger. In all cases of complete fistula the occasional escape of a little feculent matter into the passage would be amply sufficient to prevent the part healing, even if the actions of the levator and sphincter ani, and the movements of defecation, did not also interfere.

An anal fistula is at all times an annoying complaint. Even when the seat of disease is free from all inflammation and tenderness, the patient is troubled with a discharge which stains the linen, and keeps the part uncomfortably moist. The discharge is usually a thin purulent fluid; at other times it is thick, and, in complete fistula, tinged brown, from admixture of feculent matter. The discharge is more or less copious in different cases, —a circumstance depending very much on the extent of the sinuses: it also varies at different times. It occasionally becomes so thin and scanty, that the patient begins to think that the fistula is about to close, when he is disappointed by fresh irritation being set up, and the complaint becoming as annoying as ever. In complete fistula much annoyance is caused by the escape also of flatus.

Anal fistula is a disease of middle life, and occurs more frequently in men than in women. It is occasionally met with in young children, but rarely forms in advanced life, which, probably, is owing, partly, to the laxity of the rectum and sphincter in old people rendering the mucous membrane less liable to irritation and injury, and, partly, to the relief obtained by discharges from the hæmorrhoidal veins when congested.

The treatment necessary during the formation of the abscess which precedes the establishment of a fistula is rest in the recumbent posture, fomentations or the hip-bath, a poultice to the part, and mild laxatives. Leeching does not prevent suppuration taking place, and weakens the patient unnecessarily. As soon as fluctuation can be felt, the prominent or central part of the abscess should be punctured freely, to prevent the matter burrowing in the loose areolar tissue, and thus to limit the extension of the sinuses. A piece of lint should be inserted between the lips of the wound, and kept there for a few hours to prevent their closing. The local treatment must afterwards **be continued** until inflammation **has subsided,** and **the suppurating sac has** become **fistulous and indolent. An examination** may then **be made.** For this purpose I use a probe-pointed steel, or silver director, slightly curved, with **the** groove carried quite to the extremity, **and a flat** handle. The patient can be examined lying on the **side,** or leaning over a table opposite a good **light.** The director, held lightly in the hand, being inserted at the external orifice, is to be passed along the **sinus,** the oiled forefinger of the left hand being afterwards introduced into the rectum: the surgeon is then to search with care for the inner opening in the usual situation just within the sphincter. I can generally manage to detect the orifice with the point of my finger by feeling some inequality of surface, a papilla **or** slight depression in the mucous membrane. The point of the instrument having slipped **through** the opening into the rectum, comes **in contact with the** finger. The probe **should always be passed into** the fistula before **the finger is introduced into** the bowel; for **if the finger be** inserted first, the distension of **the rectum** may interfere with the operator's tracing the exact course of the fistula with the probe. It is not always easy to find the opening into the rectum. If the surgeon **fail, he must repeat** the attempt a second or a third **time, until he has** found the aperture in the **mucous** membrane, or has satisfied himself that none **exists,** being most careful to avoid using the slightest force. This is especially necessary in examining tortuous sinuses passing up the side of the rectum; for the areolar tissue yields so readily, that, without care, a passage may easily be made where none existed before. The inner opening may sometimes be detected by introducing the speculum, and exposing the mucous surface in the vicinity of the fistula, and then injecting a little chalk mixture or milk at the outer orifice. The appearance of the white fluid at a spot in the mucous membrane indicates the situation of the aperture towards which the surgeon may guide the probe. It has been proposed to inject the tincture of iodine whilst the surgeon's finger is in the rectum. The stain of iodine on the finger would indicate the site and depth of the inner opening. It may be objected that the tincture of iodine is liable to produce pain and irritation in the rectum as well as in the fistula.

The cure for anal fistula is by operation—a division of the parts intervening between inner and outer orifices, including the fibres of the external sphincter. If the fistula be recent and of slight extent the operation is not severe, so that I do not employ anæsthetics unless the patient is a sensitive, timid person, and much desires it. In long and complicated fistulæ insensibility assists the operator. The patient must be placed upon the side corresponding to the fistula. The operation is sometimes performed with the body bent over a table, but this is not so convenient. An aperient should be given, so as to obtain relief from the bowels a few hours before the operation. It is usually performed by passing a director into the rectum in the manner above described, and in a thin person its point may often be made to appear at the anus. A strong curved bistoury with a blunt extremity is then carried along the groove of the director, and the parts between the two openings are completely divided. I now generally perform the operation by passing a curved bistoury with a long probe point (fig. 7) along the fistula, and through the internal opening into the bowel, and then by depressing the handle of the instrument, I bring the point out at the anus, and, carrying the bistoury onwards, divide the intervening parts. The operation may often be facilitated by passing the forefinger of the left hand into the bowel. This is a very rapid and effectual mode of performing the operation for fistula, and gives but little pain.

Fig. 7.

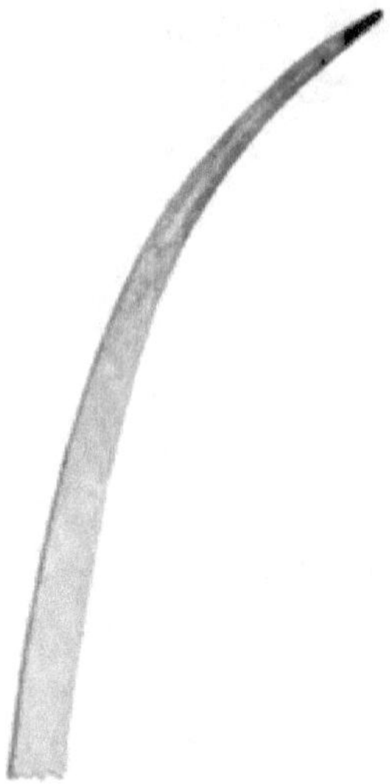

In old cases of fistula, the hardened walls of the sinus offer considerable resistance to the knife. The surgeon should not, therefore, use too slender a bistoury, for fear of the blade breaking. It is advisable, also, after the division of the parts between the two orifices of the fistula, to incise its posterior wall, which may be done with a straight bistoury or scalpel. This *reverse* incision in all cases ensures a complete division of the sphincter muscle, and a freer opening for the escape of matter, and in cases of old standing, in which the walls of the sinus are but little disposed to granulate, it tends to set up a healing action from the bottom of the wound, and to facilitate the cure. After the division of a fistula, where the undermined integuments are thin and badly nourished, it is necessary also to excise the angles or even the whole of the flaps. If this be not done the angles are apt to curl inwards, and cicatrization is often delayed or arrested. A free removal of imperfectly nourished overlapping integument much assists the cure.

After the operation a piece of damp cotton wool should be placed between the edges of the wound, and passed fully to the bottom, to prevent adhesion and to stop bleeding. A pad of cotton wool applied over this, and confined with a T bandage, will make gentle compression for a few hours. The bleeding in this operation is usually slight, and even in severe cases, where the external parts have been pretty extensively divided, the treatment recommended is quite sufficient to prevent any troublesome hæmorrhage. The bowels must be kept quiet by opiates for three or four days, and then a mild aperient will be required. The wound should be well syringed with a weak solution of Condy's fluid used tepid, and a bit of wet lint may be passed gently to the bottom of the wound after each evacuation. If the sore be slow in healing, the lint may be dipped in a slightly stimulating solution. This is the only application necessary. The wound commonly closes readily by granulation, and the functions of the sphincter are unimpaired, though in bad cases, where a very free or double division of the muscle has been necessary, the operation has been followed by weak retentive powers, causing trouble when the bowels are relaxed. In persons of feeble powers or unhealthy constitution, the wound consequent upon the operation is often very slow in healing, and if not carefully treated is apt to remain a tedious indolent sore, bringing discredit on the surgeon. The patient should not be allowed to move about, but should keep the recumbent posture pretty constantly, and, in addition to constitutional treatment, such as quinine, steel, and cod-liver oil, removal to the country or sea-side may be advisable. My rule after an operation for fistula is to watch the case until the wound is quite closed.

In those cases in which sinuses run for some distance up the side of the rectum, it was supposed, until recently, that these passages could not be obliterated without being laid open in their whole extent; and accordingly the parts were divided high up, and a severe operation performed, at the risk of serious hæmorrhage, which it was at all times difficult to arrest. In consequence of the inner opening not being sought for in the right direction, a complete fistula was often mistaken for a blind external one; and therefore, in operating, an artificial opening was made above a natural one which had escaped detection; so that not only was a larger division of parts effected than was really required, but, owing to the inner orifice below not being included in the incision, the operation not unfrequently failed. The observations of M. Ribes were consequently of great service in leading surgeons to search for the opening into the bowel near the sphincter instead of at the extremity of the fistulous sinus, and in showing that the inner orifice was present far more frequently than was commonly supposed. The improvements in the treatment of fistula which naturally sprung from these observations were early carried out, and have been strongly advocated by Mr. Syme, of Edinburgh, in his book on Diseases of the Rectum. Similar views of practice have also been enforced by the late Sir B. Brodie, in the Lecture already referred to. These eminent surgeons also consider that, when a fistula passes for some distance upwards along the side of the rectum, it is not necessary that it should be divided in its whole extent; and that, if the parts intervening between the inner and outer openings below be freely cut through, the sinus above will probably close, and the patient be cured by a simple and slight operation. My own experience does not enable me to coincide altogether in these views, for in many cases I have found that the sinus running up the rectum will not close without a freer division of parts. The wound below is apt to assume an unhealthy aspect, and a free discharge continues. In a case of this kind, with a sinus burrowing close to the mucous membrane, I have sometimes passed up to its extremity a straight director, and carried along its groove one of the blunt-pointed blades of a pair of knife-cutting scissors, whilst the other has ascended the rectum, and, by closing them, have divided the intervening membrane and exposed the sinus. I have then lodged a tuft of cotton wool in the gap, which has afterwards healed in the usual way. This is a simple and easy operation, attended with but little risk of serious hæmorrhage even when the barrier is divided high up, for the sinus in burrowing close to the mucous membrane detaches it from the chief vessels and parts beneath. A fistula sometimes, however, penetrates into the areolar tissue of the pelvis outside the rectum, having a thick barrier between the sinus and the interior of the bowel. The operation just described would then be attended with danger, and the case must be treated differently.

When the opening into the rectum is more than an inch and a half above the external sphincter, or when the sinus penetrates to this or a greater distance in the areolar tissue outside the bowel, the division cannot be made without risk of hæmorrhage, which the surgeon may find difficulty in arresting—indeed, death from bleeding has happened after the division of a rectal fistula high up. The hæmorrhage may go on without the patient being aware of it, no blood escaping externally to cause alarm. He feels perhaps weak or faint, experiences uneasiness in the rectum, and a sense of fulness which at length obliges him to go to stool, and then he passes a quantity of blood which has gradually accumulated in the bowel. The best mode of stopping bleeding from a deep wound at the anus extending into the rectum is to plug it with fine cotton wool. Some small pieces, previously damped, the water being well pressed out, should be lodged in the bottom of the wound with a probe. The wool must on no account be oiled, or it will slip out and be of no use. To maintain deep-seated pressure tufts of wool should also be applied over the anus, and covered with some pieces of lint, all which should be kept in place with a T bandage applied somewhat firmly. The plugs need not be disturbed for several days, and means must be taken to keep the bowels confined for that period. The wound can be effectually plugged with sponge, but this substance is liable to adhere to the parts, and to be difficult of removal.

Cases of the kind just described may be treated by ligature. The ligature, if properly applied, answers very well, and is less tedious and painful than is commonly supposed. Its application to ordinary cases of fistula, though often practised formerly, is now seldom resorted to, the knife being found a less painful and tedious mode of curing the disease. Attempts have indeed been made recently by Professor Dittel, of Vienna, and Mr. Allingham, to revive the treatment by ligature by substituting elastic india-rubber cord for silk or twine. The continued tension obtainable from india-rubber I have long used, in removing ligatures from stumps after amputation, from nævi as well as from fistulæ, by attaching an india-rubber ring to the end of the silk ligature, and rendering this tense by securing the stretched ring to some convenient fixed point near the part ligatured. The india-rubber cord, when tightened, exerts more complete pressure, but the pressure is not so easily increased when required as with a ring. The knife is, however, the best remedy for ordinary fistula, and the elastic ligature may be reserved for rare and special cases, such as I have alluded to. The passage of the ligature when the sinus runs high up by the side of the bowel is not an easy matter, and requires some manipulative skill. Various ingenious instruments have been contrived to facilitate the proceeding. That which I have found very convenient is a fine long flat silver tube, slightly curved, fixed in a handle, which can be passed along the fistula into the rectum. If no internal orifice exist, a sharp trocar-needle can be passed through the canula, and carried with the tube through the mucous membrane into the bowel.

Metallic wire can then be introduced through the tube into the rectum to a sufficient extent to admit of the surgeon seizing it with a forceps, or looping his finger with it, and drawing it out at the anus. The tube having been withdrawn, a piece of strong twine attached to the wire can be drawn through the fistula. The two ends of the ligature must then be secured to a strong india-rubber ring, which can be kept on the stretch until the parts are cut through, and the ligature comes away. The tube is well suited to facilitate the passage of platinum wire, by which means the galvanic cautery can be applied to the immediate division of a fistula, in persons objecting to the knife. The ligature was adopted with success in the following case :—

Chronic sinuses and fistula high up, cured by incisions and ligature.

Case 17.—In March, 1862, I saw, with Mr. Smart of Hackney, a gentleman of middle age, who had been troubled for several months with obstinate sinuses in the buttock opening a long distance from the anus. He had been under the care of a hospital surgeon, who had not considered the sinuses as connected in any way with the rectum, and had treated them chiefly by iodine injections, but without success. Chloroform having been given, I freely laid open the branching sinuses, cutting through a great thickness of parts, and was then able to track one passing off at a right angle to the side of the rectum. I divided it into the bowel, cutting through the sphincter muscle. A sinus running some distance up close to the rectum was not interfered with. The patient was afterwards kept in bed, the sinuses laid open were dressed from the bottom, and the general health was well attended to. In about four weeks the wounds in the buttock gradually healed, but there was a constant discharge from the sinus near the bowel. I divided this into the rectum with the director and scissors the distance of an inch, but was afraid to go higher, as I felt a large artery in the way. As the discharge still continued, after a fortnight, I passed a curved silver tube quite to the upper part of the sinus and forced it into the bowel. I managed with some trouble to carry a silver wire with twine attached through the canula into the rectum, and to bring it out externally. The ligature was properly secured to a screw tourniquet, by which means the intervening parts were quickly cut through. This did well, but the confinement and treatment began to affect his general health; and as a disposition to the formation of a sinus in the direction of the coccyx manifested itself, we sent him in June to the sea-side, where he gradually recovered his health, and the parts completely healed.

In cases of blind external fistula, in which the surgeon is satisfied, by a sufficient examination, that there is no internal orifice, the probe point of the fistula-knife, or the point of the director should be carried to that part of the sinus which is close to the mucous membrane of the rectum, and made to bear steadily against the end of the finger until the membrane is perforated, care being taken that sufficient support is given by the finger to prevent the bowel being in any degree detached from the neighbouring structures by the pressure of the director. The intervening parts can then be divided, as in complete fistula. The spot where the membrane is denuded will generally be found a short distance only above the external sphincter. We read in books of *blind internal fistula*, in which an opening into the bowel leads to a fistula without any external orifice. Such cases are but rarely met with in practice. The external opening sometimes closes for a short time, the spot being indicated by redness and induration; but sooner or later it re-opens, and the discharge returns, or a fresh opening is formed at some little distance off. It may happen, however, that the original ulcerated opening in the rectum being large, the matter from the abscess formed in the areolar tissue finds its way so readily into the bowel that the abscess does not burrow towards the surface. This is not a common case. The situation of the suppurating cavity may be ascertained externally by a sort of hollow, or indistinct fluctuating feel. When this is the case, a bistoury plunged into the sac will render the fistula complete, and it may then be treated in the usual manner.

Case 18.—A lady whom I attended, with Mr. Cuolahan of Bermondsey, had a suppurating cavity with a large internal opening at the lower part of the bowel. This cavity passed some little distance up the rectum, causing an internal projection of the mucous surface. In addition to a fetid purulent discharge there was great pain in defecation and at other times. After an external incision through the sphincter, carried into the internal opening, I divided the mucous coat of the bowel forming the side of the cavity with the scissors carried along a director in the way already described. The parts gradually healed up afterwards, and the result was quite satisfactory.

A blind internal fistula especially when deeply seated is liable to be overlooked. I attended with Dr. Priestley a lady who had this complaint attended with so copious a discharge that she was emaciated and almost worn out by its long continuance. She had been attended by several practitioners, but the nature of the case had escaped detection for several years. A lady with a similar fistula who consulted me, informed me that the discharge, which was abundant and kept her linen constantly soiled, had been referred to the vagina, and different kinds of treatment had been adopted without reference to the real source of the discharge notwithstanding her complaints of uneasiness experienced in the rectum.

The sinus communicating with the rectum does not always correspond with the outer orifice, but opens into a sinus in the buttock passing at an angle to it, so that a probe introduced at the external opening traverses the buttock, but it cannot well be made to enter the upper sinus. In such a case the surgeon is liable to conclude that the fistula is confined to the buttock, and does not implicate the rectum. This mistake was made in Case 17. An incision laying open the lower sinus will expose the entrance of the one running up to the bowel, and set the surgeon right. In cases where the matter burrows in the buttock, and comes to the surface some two or three inches or more from the anus, it is not always necessary to lay this sinus open for its whole extent, which would be a severe operation. Using the probe end of the director as a guide, the surgeon may make an external artificial opening into that part of the fistula which is near the anus, and then divide the structures between this orifice and the internal one in the usual way, by which means, the communication with the outer part of the fistula being cut off, it closes without difficulty, whilst the internal wound heals by granulation from the bottom.

It is difficult to describe the treatment required in the different forms of complicated fistula; so much depends on the peculiarity of each case, no two being ever exactly alike. In the *horse-shoe* fistula, a free opening into the rectum on one side will sometimes be sufficient, the outer opening of the sinus on the other side being dilated so as to allow a free escape of any **pent-up matter.** In several instances, however, I have had **occasion** to perform **the** operation on both sides, and with **perfect success in curing the sinuses.** After the double **division of the sphincter the patient may be unable to retain liquid fæces; but this need not cause uneasiness, for** I have always found that **when** the wounds healed, **and sometimes** even before, the functions of the muscle **were restored, and** no permanent inconvenience resulted **from its division on** the two **sides.** When there **are two** internal orifices on the same side, it is desirable, if possible, to include both of them in the incision, or, **after the upper one** has been divided in the usual manner, to extend the incision so as to make the **lower** one open into it. If the interval between the two inner openings be great, one being situated high up, it will be as well to operate from the lower one, after which the free passage of the fæces and inactive state of the sphincter may allow of the upper **opening and sinus closing,** though this cannot be **at all relied on. When an inner opening exists on both sides of the rectum, with only one external orifice, the usual operation may be performed at the side** on which the outer opening is situated, taking the chance of the other internal aperture closing spontaneously. Should this not take place, a second operation can afterwards be done on the other side. In these complicated cases the condition of the patient's health **often** precludes the performance of any operation.

Some difference of **opinion exists** respecting the

propriety of operating for fistula on persons affected with phthisis, those opposed to the use of the knife contending, that the discharge from the fistula, like an issue, is beneficial in arresting the progress of the disease in the lungs; whilst others maintain, that the irritation and suppuration are injurious in adding to the patient's exhaustion. In advanced cases of phthisis no judicious surgeon would venture to use the knife, but I am convinced that in an early stage of the disease a source of debility may often be removed, and the comfort of the patient promoted by an operation, which I have performed with benefit in several instances. The wound, however, is usually very slow in healing, and continued attention in the application of stimulating lotions and ointments to the wound may be required to obtain a satisfactory result.

A fistula connected with a carious state of the ischium or sacrum is unfit for operation, unless the surgeon can reach the diseased bone, and, if necessary, gouge it. Sinuses in the perineum are sometimes found to open into the rectum as well as into the urethra, and the communication allows the escape of gas, and sometimes of thin feculent matter, into the urinary passage, to the great annoyance and distress of the patient. These sinuses originate in inflammation, and abscess of the prostate gland, and do not belong, exactly, to the diseases of the rectum. To obtain a cure, the fistulous passages require to be freely laid open into the bowel, and the wound must afterwards be dressed from the bottom with lint or cotton wool. The outer orifice of a simple fistula in ano is sometimes seated in the perineum so far in front of the anus, and directly in the course of the urethra, as to lead to the suspicion of its being an urethral instead of a rectal fistula. But as no urine escapes from the orifice when the patient makes water, and as a probe takes the direction of the anus, the nature of the case is easily ascertained, and after laying open the fistulous passage in the perineum, the surgeon is able to trace the sinus leading into the rectum.

Some years back I was consulted by a young married woman who had not only a fistula which opened by the side of the rectum, but another which communicated with the vagina, and a third that opened at the lower part of the labium. So much feculent matter passed into the vagina that it was evident the communication between it and the rectum was pretty free. Though in a miserable condition, she would not consent to undergo an operation, and I lost sight of the case. In a case of this kind, if both the sphincter ani and the sphincter vaginæ are divided, it is found that the patient loses the power of retaining her fæces. The late Sir B. Brodie stated, that a lady consulted him with a fistula communicating with the rectum in front, and opening externally just at the beginning of the vagina. He merely made a free division of the sphincter ani on both sides, so as to set it completely at liberty. The discharge from the fistula gradually diminished, and, some five months after the operation, the fistula appeared soundly healed. I intended to have adopted a somewhat similar plan in the case above alluded to, but the large size of the opening into the vagina would, I expect, have rendered some further proceeding necessary.

CHAPTER X.

CATARRH OF THE RECTUM.

A mucous discharge from the rectum, of the character of catarrh, is a rare complaint. It occurs like other catarrhs from a chill. The symptoms consist of a slight diarrhœa and a frequent pressing desire to relieve the bowel where only a colourless mucus is passed. This is attended with an uneasy sensation in the back of the sacrum. The complaint is to be treated by rest, warm hip baths, and small doses of Dover's powder. It then passes off in a few days. I have a patient of marked gouty tendency who has been troubled with slight attacks of this complaint, and has also suffered at other times from attacks of cystitis after chills.

I have been consulted for purulent discharges from the rectum supposed to arise from an internal fistula difficult of detection, and I have discovered that the pus came from the mucous surface of the bowel. I have had my suspicion that the affection was communicated. I attended with Dr. Gueneau de Mussy and Mr. Aikin a young married foreign lady, with a copious purulent rectal discharge unattended with pain. It was treated by astringent injections, and the mucous membrane was freely brushed over with a strong solution of nitrate of silver through a speculum. The complaint proved obstinate and difficult of cure.

CHAPTER XI.

CHRONIC ULCERATION OF THE RECTUM.

INQUIRIES into the morbid anatomy of the rectum have brought under my notice many instances of ulceration of its mucous lining, not only in cases of dysentery, and as a consequence of the ordinary diseases of the part, such as stricture and cancer, but sometimes as a separate affection. In several specimens which I have examined, ulceration was diffused over a considerable extent of surface. I have observed the lower part of the rectum quite stripped of its mucous membrane for a distance of two or three inches. This extensive disease is sometimes, indeed generally, attended with thickening and consolidation of the subjacent tissues, without diminution in the calibre of the bowel. The muscular coat is, in some instances, hypertrophied. In one case, the mucous coat for a short distance within the sphincter was so riddled with holes as to form, as it is described in the post-mortem book, "a perfect cribriform tissue," the submucous tissues being at the same time much thickened. I have seen the mucous membrane ulcerated in patches, the sound portions being in some places detached from the muscular fibres beneath, so as to form bridges more or less broad, or merely some narrow bands or bridles. There were, frequently, abscesses in the thickened tissues around the diseased rectum, and fistulous passages opening externally. In two instances, ulceration had produced a perforated opening communicating with the peritoneum, death having been caused by the escape of some feculent matter into the abdomen, and inflammation of the serous membrane. In other cases the peritoneum was involved in the consolidation, and inflamed without being perforated, the omentum in one case being adherent to the anterior part of the rectum.

The history of these cases of extensive ulceration was not always sufficiently clear to enable me to trace the origin of the disease satisfactorily; but in numerous instances it resulted from dysentery, many of the sufferers having recently returned from warm climates, or been exposed to hardships at sea. They were all cases of a chronic character, the morbid parts having been taken from subjects who had suffered for a long period from a complaint of the lower bowel. In a few of the cases it seemed probable, from what could be gathered of the history, that there had been chronic inflammation of the coats of the rectum, and ulceration, which had been aggravated, if not produced, by the improper and rough use of bougies for some slight or supposed contraction of the passage. Some years ago I had a woman, aged seventy-nine, under my care, who had unequivocal symptoms of ulceration high up in the rectum. She had suffered from disease of this part for more than fifteen years, and about eight years before had been treated for stricture by a hospital surgeon, who occasionally passed bougies for upwards of two years; yet, upon a recent examination, I could detect no contraction in the bowel, but the mucous surface high up felt rough and irregular. There can be no doubt that, in a rectum rendered irritable by drastic purgatives, or acrid secretions and evacuations, abrasion of the mucous surface by a clyster-pipe or indurated fæces would be sufficient to excite ulceration and generate chronic inflammation.

When we consider the great frequency of ulceration in the mucous structures at the commencement of the alimentary canal, in constitutional syphilis, as in the mouth and throat, we cannot be surprised to find that similar changes sometimes take place in the mucous membrane at the termination of this passage. Some years ago, the late Mr. Avery exhibited at the Pathological Society a specimen of ulceration of the rectum, the history of which clearly showed the connexion of the lesion with syphilis. Immediately within the anus, which was surrounded by a circle of vegetations, the ulcer commenced, extending three inches upwards, and occupying the whole of the internal surface of the rectum to that extent. The edges were rough and uneven above, and below soft and rounded; the whole surface was smooth, exhibiting the muscular fibres of the intestine quite bare. The patient, a young woman, aged twenty-two, died in the Charing Cross Hospital from erysipelas of the face, and had been troubled with a discharge from the rectum for about seven months previously. She had been in the hospital a year and a half before with an extensive sloughing ulcer in the fourchette. When she died she had numerous indelible marks of syphilitic eruption on the limbs and trunk, and was suffering from sore-throat.[9]

It is a question whether syphilitic ulceration in the rectum is due to direct contagion, the lining membrane of the bowel becoming inoculated with

---

[9] Trans. Pathological Society, vol. i. p. 94.

the secretion from the sores on the vulva or in a more direct way, or whether the ulceration is a result of constitutional disease. Some of the best authorities of the present day who have directed attention to the subject, Gosselin [1] and Lancereaux of Paris, Drs. Bumstead [2] and Mason [3] of New York, regard it as a local sore, and not as a symptom of general infection. This view is supported by the circumstance that the ulceration has been observed almost invariably in women, though Gosselin in some rare instances found it in the male. Cases of the disease have not fallen under my notice in recent years, but formerly when actively engaged in hospital practice several cases of ulceration of the bowel causing stricture, which I was induced at the time to refer to constitutional syphilis, came under my care. Syphilitic ulcers are usually large in size, and often involve the deeper structures of the coats of the rectum; the result is, that the healing process is very apt to cause a serious contraction in the passage.

I have stated in a previous chapter, that rectal fistula sometimes arises from tubercular ulceration in the bowel, and that in these cases the inner orifice of the sinus is somewhat large in size. The ulcer, instead of penetrating into the submucous tissues and causing fistula, may spread superficially, and give rise to a troublesome chronic sore in the rectum. In the cases which I have met with the ulcer was nearly always small in size, but often tedious, and

---

[1] Archives Générales de Méd. 1854.
[2] Pathology and Treatment of Venereal Diseases, 3rd edit.
[3] American Journal of Medical Sciences, vol. 65. 1873.

indisposed to heal. The patients were evidently strumous subjects, and a man, aged thirty-five, under my care in the London Hospital, with an ulcer in the rectum the size of a florin, had well-marked symptoms of phthisis.

The chief symptoms which may be considered as referable to chronic ulceration of the rectum are—a purulent discharge from the anus, more or less copious; motions generally loose, and mixed or coated with a slimy fluid, and streaked with blood; soreness in passing stools, and occasionally tenesmus. The pain in defecation varies considerably, being in some cases severe, and in others very trifling. Indeed, it is surprising how little suffering is often caused by the actions of the rectum and passage of the fæces, in cases of large ulceration of the mucous surface. The old lady, to whose case I have briefly alluded, had very little uneasiness in passing her stools; and Mr. Avery's patient, though affected with extensive disease, suffered very little pain, but she was troubled with a copious discharge. The suffering of the patient much depends on the position of the ulcer. Whether large or small, if it extends low down, so as to come within the grasp of the sphincter muscle, the pain is generally severe and persistent after defecation; and, in addition to other treatment, an incision through the lower margin of the ulcer is often required, to release it from the action of the sphincter.

The characters, position, and extent of chronic ulceration in the rectum must be ascertained by examination with the finger and with the speculum. The surgeon will be able to feel a rough uneven surface, more or less indented or depressed, and frequently hardness and consolidation of the walls of the rectum. The appearance of the sore, in the lower part of the bowel, may be seen through the speculum, and this instrument is most useful for the application of local remedies.

The treatment suitable to chronic ulceration greatly depends on the nature and extent of the disease, and upon the constitutional condition of the patient. In severe cases I always keep the patient at rest in the recumbent posture. In extensive destruction of the mucous surface, attended with frequent discharges, especially when the disease originates in dysentery, vegetable astringents, such as simaruba and krameria, combined with the mineral acids and opiates, are generally of great service in restraining the tenesmus and irritating evacuations and discharges. The nitrate of bismuth, with magnesia and anodynes, also affords great relief, and the sulphate of copper with opium may often be given with advantage. When the ulceration is consequent on constitutional syphilis or scrofula, the remedies appropriate to these diseases are required. The diet must be carefully regulated. The local treatment consists in the application of weak solutions of nitrate of silver, and anodyne injections with mucilage, or anodyne suppositories.

The following cases will serve to illustrate the preceding observations :—

Chronic tubercular ulcer in the rectum, cured.

Case 19.—Capt. M——, aged thirty-five, of strumous appearance, a sea-captain in the merchant service, consulted me in May, 1856, on account of a complaint in the rectum, from which he had suffered more or less for eighteen months. He had been a good deal in warm climates, had been exposed to hardships in his occupation, and had deposits, which I regarded as of a tubercular character, in both testicles. He complained of a severe scalding pain which lasted about ten minutes after an evacuation, being more severe when it was solid. He also passed a little blood with his motions, and had constantly a slight discharge. On examination I detected at a short distance within the bowel an ulcer with indurated edges. It was nearly the size of a florin, and its lower part was within the circle of the sphincter. I made a slight incision of the lower part of the sore under chloroform. I kept him afterwards in the recumbent posture, gave him decoction of sarsaparilla with the iodide of potassium, and electuary of senna at bed-time. With the exception of a slight smarting sensation on the two first occasions of a relief from the bowels, the uneasiness after stool entirely ceased. At the end of a week cotton wool moistened with a weak solution of nitrate of silver was applied to the sore. In a month it had quite healed, and the patient sailed for Australia, and I have since heard that he has remained well.

Chronic strumous ulcer in the rectum, difficult of cure.

Case 20.—A clergyman from the country, aged fifty-two, a tall spare man, of pallid countenance and strumous habit, suffering from an affection of the rectum, consulted me in March, 1859. He stated, that he first noticed uneasiness in defecation, with occasional bleeding, four months previously. The pain was increasing, and was experienced after exercise, and he was sometimes troubled with an urgent desire to pass his motions, especially in the morning and at night. He also complained of occasional pains extending down the back of the left thigh. The blood which passed was very small in quantity. The complaint occurred without any apparent cause. He had been under treatment, but without getting relief. On examination with the finger I readily detected a small oval ulcer towards the back and left side of the rectum, about an inch from the anus, and behind the sphincter. The sore was quite superficial, and about a quarter of an inch in diameter. With the speculum it was seen to present a pale aspect, which contrasted with the tumid and deep red mucous membrane around. It bled readily, even when touched gently with a camel-hair brush. He was kept at rest, chiefly in the recumbent posture, and allowed to take carriage exercise only. The motions were kept soft by a table-spoonful of olive oil taken at bed-time, and a weak solution of nitrate of silver was applied to the sore, through a speculum, two or three times a week. In a fortnight there was great improvement. The pain had diminished, and the bleeding occurred much less frequently, but still the sore did not heal. The sulphate of copper in substance was applied to it, and some tonic medicine prescribed, but the sore remained stationary. The patient began to complain of more pain in the part, and down the thigh, and was getting depressed. Sir James Paget saw the case in consultation, and we decided on applying to the ulcer the acid nitrate of mercury. It caused a burning sensation, but not severe. The patient was kept strictly at rest, and the bowels were prevented acting for two days. Some little progress in healing then took place, but as the sore had not closed at the end of four weeks, the caustic application was repeated. Under this treatment the ulcer at length healed, and the patient returned to his duties in the country. About nine months afterwards he came up to see me, in consequence of a slight return of the bleeding, and of pain down the right thigh. On examination I found a slight superficial ulcer, or rather abrasion, at the site of the old sore. A solution of nitrate of silver was applied, and with rest, nitro-muriatic acid, and cod-liver oil, the sore soon healed, and the part has since remained well.

Extensive ulceration of the rectum cured, but followed by contraction from cicatrization.

Case 21.—Hannah L——, aged twenty-five, a sailor's wife, was admitted into the London Hospital, December 12, 1860. About six months ago she began to feel a constant bearing down **pain in the lower** bowel, and to pass blood in her motions. The stools were frequent, of a dark slimy character, and attended with severe pain. They always contained blood, either fluid or in clots. She constantly suffered more or less pain, but was easier **in bed** than when up. Her pulse was rapid and feeble, her appetite bad, and her tongue clean, but **very** florid. She was pale and anæmic, and had **an anxious expression. There were scars in the** neck from **strumous disease of the glands. She stated that she had been married two years, and since then had menstruated only** once or twice, but had had no family. **She had been troubled with piles, which had** bled occasionally for five or six years. There were **some flat** cutaneous growths around the anus. On passing my finger, I discovered extensive ulcera**tion in the bowel, an irregular surface** in parts rugged **and hard,** at some points exquisitely **tender.** The finger when **withdrawn** was tinged with blood. I ordered her to remain **in bed, a** solution of nitrate **of silver** (gr. ij.—ʒj.) to be **injected** into the bowel every morning, a mixture containing bismuth mucilage and Battley's solution to be taken three times a day, and a suppository **of** soap and opium to be passed every night. Dec. 20, the pain had diminished, the bowels had become less relaxed, and her general condition had improved. The suppositories were of much service in relieving the pain and obtaining rest. The blood in the motions **also** soon afterwards ceased. A month after she had **been under treatment I** found on examination that the **surface of the bowel was soft and smooth, and had lost all tenderness, but I discovered about two inches from the anus a** distinct **contraction in the bowel. This, which** resulted from the cicatrization **of** the ulcer, was **counteracted** by the passage of a No. 8 bougie on alternate days. The muriated tincture of iron was now given instead of the bismuth mixture, and she was allowed to leave her bed, and walk **in the** hospital garden. **Under** this treatment the contraction disappeared. All mucous discharge ceased, she regained **her** health and strength, and was discharged cured, March 1st, 1862.

Syphilitic ulceration of the rectum cured, but terminating in a close stricture of the gut.

Case 22.—E. G., aged twenty-nine, a married woman, who had never borne children, was admitted into the hospital, November 3, 1857, on account of a stricture in the rectum. It appeared that she had been seduced when young, had since led a profligate life, and had early contracted syphilis. She had latterly been married to a sailor. About two years previously she first noticed a complaint in the rectum, and in consequence of pain and straining in passing her motions, and of a slimy discharge tinged with blood, she was admitted into the hospital under my care in November, 1856. At that time I viewed the case as one of secondary syphilitic ulceration of the rectum. There was no obstruction in the passage, but the mucous membrane of the lower part of the bowel was extensively ulcerated. She had, also, painful nodes on the forehead and on both shin-bones, besides other marks of undoubted syphilis. She was, too, in impaired health. Under treatment, consisting chiefly of decoction of sarsaparilla with iodide of potassium, opium, rest in bed, and mild stimulating applications to the ulcer in the rectum, she much improved in health, the pains in the bones subsided, the irritation in and discharge from the lower bowel diminished, she passed her motions without pain, and, on examination, the rectal ulcer was found to be healing. I noticed at that time the circumstance of the healing being attended with a commencing stricture in the rectum. Her husband having returned from sea and having sent for her, she was discharged at her own request in the following December, after a residence of five weeks in the hospital, and was made an out-patient. She neglected, however, to attend, and I saw nothing more of her until November 3, 1857, when she again applied for admission on account of great difficulty in passing her motions. She was looking in better health than when she left the hospital, having gained flesh. There were two small copper-coloured patches on the face, some indistinct nodes on the shin-bones, and one also on the forehead, and she was subject to occasional pains in these parts. On examination I found about an inch and a half from the anus an extremely tight stricture of the rectum. The opening was so contracted, that it would admit only the smallest-sized rectum bougie. There was also a complete fistula in ano below the stricture, and a sinus some three inches in length running up between the rectum and vagina, opening into the latter near its orifice, but without communicating with the bowel. The sinus seemed to terminate in a blind extremity. The treatment consisted in small doses of castor oil; the passage of a bougie every other day, the instrument being retained for an hour; compound decoction of sarsaparilla with five grains of iodide of potassium three times a day, and iodine paint to the nodes. A few days after her admission the catamenia appeared for the first time for two years. Mr. Ryott, the dresser, very gradually increased the size of the bougies, and on the 24th succeeded in passing No. 4, but not without pain and the escape of a little blood. On the 27th it passed very easily, and without any bleeding. I had intended dividing the slight fistula, but on the 2nd of December she obtained leave to go out to obtain part of her husband's pay, and did not return to the hospital, and I have since lost sight of her.

The margin of the anus and lower part of the rectum are subject to an intractable form of ulceration, which slowly creeps on until it causes an extensive destruction of parts. The ulcer is superficial and similar in character to the sore which occasionally attacks the face, known by the term *rodent*. It is difficult of cure, and after the sore has healed ulceration is very apt to recur, but the disease is not malignant, and there is an absence of induration and of enlargement of inguinal glands, by which it may be distinguished from epithelial cancer. The following well-marked case recently came under my notice:—

Intractable rodent ulcer of the anus and rectum.

Case 23.—An aged member of the medical profession was troubled with an obstinate ulcer at the anus for many years. It nearly healed up twice, but never quite, and the ulceration returned and gradually spread. When this gentleman first consulted me, I found a large sore involving the skin at the side of the anus and the mucous membrane, extending some way up the bowel, and furnishing a thin discharge. Some of his professional friends supposed that this intractable disease was malignant; but there was no induration, or evidence of adventitious deposit, and no sharp darting pains. Under applications of a solution of nitrate of silver, rest in the recumbent posture, and steel medicine, the sore nearly healed, without any contraction of the part; but the patient afterwards neglected himself, the ulceration returned and spread until it formed an immense sore, which continued to increase for many months, until he died at the advanced age of eighty-one.

CHAPTER XII.

STRICTURE OF THE RECTUM.

The rectum, like other mucous canals,—as the œsophagus and urethra,—is liable to obstruction from a contraction of its walls, forming the disease called *stricture*. The contraction may be very limited in extent, and the stricture is then termed *annular;* or the contraction may include a portion, more or less considerable, of the coats of the bowel. On anatomical examination, the mucous membrane involved in the contraction is found tumid, thickened, and congested. This membrane is closely adherent;

and, when carefully dissected off, the submucous areolar tissue is observed to be condensed—to consist of close-set fibrous tissue, sometimes for a limited extent, as in annular stricture, where it surrounds the gut, and lengthwise gradually blends with the healthy areolar tissue above and below, but more frequently forming a callous indurated mass from half an inch to two or more in length. The thickening may be confined to part only of the circumference of the rectum, or may be greater on one side than on the other, contracting the canal irregularly, and forming a winding passage: or the induration may involve the greater part or the whole of the gut. In several specimens of the disease which I have examined, the calibre of the rectum was thus diminished in various degrees. In the Museum of King's College there is a preparation showing great thickening and consolidation of the entire walls of the rectum, with hypertrophy of the muscular coat, and considerable narrowing of the passage. The peritoneum investing the contracted bowel generally retains its healthy structure and appearance. Writers have described a form of stricture of the rectum produced by bands stretching across the canal. No instance of the kind has fallen under my observation; and I suspect that the bands were merely broad folds of the mucous membrane, which were supposed to intercept the passage.

Above the stricture the rectum is commonly dilated and thickened. The enlargement results not from a yielding of the intestine, but a general hypertrophy of the bowel, and particularly of the muscular coat, the fibres of which are remarkably large and distinct. The mucous membrane at this part is rarely healthy. It is generally red from capillary injection, and extensively eroded or ulcerated, the diseased surface supplying, during life, a purulent discharge. There may be ulcerated apertures leading to fistulous passages which extend for some distance, and open externally near the anus, or as far off as the buttock. In the body of a woman who had a close stricture in the rectum an inch from the anus, in addition to a large ulcer in the bowel above the contraction, I found a fistula communicating with the vagina. There may be but little alteration in the bowel below the stricture, but it is generally in some way diseased. There is frequently diffuse ulceration of the mucous membrane, and sometimes hæmorrhoids, or a complete fistula in the usual situation near the anus. Sinuses may exist burrowing in different directions amongst the thickened tissues around the lower part of the bowel.

Some difference of opinion exists respecting the usual seat of a stricture in the rectum. It varies, but is commonly at the lower part of the gut from an inch and a half to two inches from the anus, and easily within reach of the finger. In a table of twenty-eight cases under my care in public and private practice, in twenty-one the stricture was at this distance. In two it was somewhat nearer the anus, and in five at a greater distance. In three of the latter the stricture was at the point where the sigmoid flexure terminates in the rectum, which naturally presents a slight contraction, and is well known to be liable to stricture.

I have twice met with two distinct strictures in

the rectum (Cases 24 and 27). Such cases are rare. One originated in dysentery.

The pathological changes causing stricture, originate in chronic inflammation of the mucous and submucous areolar tissue of the rectum, either limited in extent, or affecting the greater part of the intestine. It is often impossible to fix on the exciting cause in a particular case; but the part is exposed to so many sources of irritation, from unhealthy and acrid secretions, the lodgment and passage of hardened fæces, injuries from foreign bodies, as fishbones, &c., the disturbance produced by undue muscular action in tenesmus and forcible defecation, that the occurrence of a slow inflammation of the coats, ending in contraction, cannot be viewed with surprise. Women, in whom the disease is much more common than in men, have sometimes ascribed its origin to a difficult labour, by which, there can be no doubt, injury may be inflicted on the bowel, so as to lay the foundation for chronic disease. In the table of twenty-eight cases above referred to, twenty were women, and in nine of them the disease commenced after labour, in some instances being distinctly attributed to an injury at that time. In a case of close stricture quite at the lower part of the rectum in a woman under my care in the London Hospital, the disease originated in an injury of the part from a severe kick inflicted four years previously. At page 147, I have related the case of a girl, aged fourteen, with stricture of the rectum, caused by a fall on the brass button of a perambulator. There is a preparation of much interest in the Museum of St. Bartholomew's Hospital, taken from a child five years old. Ten months before death, in the endeavour to administer an enema, a clyster-pipe was forced through the adjacent walls of the rectum and vagina. At the part thus injured there is a small depression in the wall of the vagina, and a long, pale, and irregular cicatrix in that of the rectum. Near this cicatrix, also, there are traces of small healed ulcers of the mucous membrane of the rectum. Just below the cicatrix, at a distance of about an inch from the anus, the canal of the rectum is reduced to an eighth of an inch in diameter, and the adjacent tissues are indurated. Above this stricture the intestine is greatly dilated.

Although strictures of the rectum are generally produced, as I have described, by chronic inflammation chiefly of the submucous areolar tissue, without any breach in the mucous surface, they also originate in another way—viz. in the contraction consequent upon the healing of ulcers or wounds in the bowel; and I believe that this takes place more frequently than is generally supposed. That ulceration of the rectum is a common disease, I have shown in the preceding chapter; and it might be expected that sores of any considerable extent and depth would be very liable to produce in the process of healing more or less contraction, just as we observe in repairs of breaches of surface in the skin. It appears, however, that a superficial ulcer in the alimentary canal, or one limited to the mucous membrane, heals without giving rise to any diminution of the passage. This is well seen in the cicatrization of the typhoid ulcer, in which the muscular coat forms the base of the sore. I have also observed healed ulcers without any puckering or contraction in the stomach. After attacks of dysentery extensive ulcers in the colon very often become cicatrized in this way. But, in the dysenteric and phagædenic venereal ulcers, the destructive process is liable to penetrate deeper, destroying the muscular structure, and reaching the fibro-areolar tissue beneath. When this occurs, the parts around are drawn in by the process of cicatrization to fill up the gap, and the passage of the bowel becomes contracted. Two cases, in which chronic ulcer in the rectum had been followed by contraction to such a degree as to prove fatal, have been recorded by Cruveilhier.[4]

Ulceration of the rectum from chronic dysentery, terminating in a double stricture of the gut.

Case 24.—W. H ——, aged thirty-one, a seaman in Her Majesty's navy, was admitted into the London Hospital, July 16, 1857, having recently been discharged from the service invalided. He had served in the Black Sea and in the Crimea during the Russian war the year before, and had been much exposed to severe weather, and at times had been unable to obtain good and sufficient nourishment. Under these privations he suffered severely from dysentery. He recovered partially from the attack, but did not regain his health, and his bowels never acted properly, his motions being small in size, slimy, and passing with pain and forcing, and he was continually obliged to take aperient medicine. His complexion was clear, but he was very weak, and much emaciated. On examination I found a slight fistula near the anus, and the interior of the rectum partly abraded and irregular from ulceration, which extended some distance up the bowel. About an inch and a half from the anus there was a distinct contraction in the gut, not very close nor firm, and apparently of recent formation. The point of my forefinger could be passed into it. Being of opinion that all active disease in the

[4] Anatomie Pathologique, livraison xxv.

bowel had ceased, and that the contraction resulted from the advancing cicatrization of the ulcer, I put the patient upon a nutritious diet, ordered him to take small doses of castor oil to soften the motions, and afterwards prescribed a bitter infusion, with nitro-muriatic acid. I at once commenced the use of bougies, introducing a small size at first, and directing the dresser to pass one every other day, and gradually to increase its size. The fistula was divided, and the wound healed readily. The stricture in the urethra was also dilated until it admitted a No. 8 catheter. In the month of August, after the passage of the instrument into the urethra by the dresser, accompanied with pain and difficulty, an abscess formed in the perineum. This was opened, and the wound healed very readily. During September I was away from London; on my return at the end of the month I found that my patient had continued to improve in health, had gained flesh and strength, and was suffering very little from his rectum. The bougie was still passed occasionally. Early in October it was observed that his motions diminished in size, and he was soon obliged to strain in passing them, and had frequent calls to the closet. I directed the dresser to pass a long tube, and an injection to be given. In doing so he discovered a stricture considerably above the original contraction, and called my attention to it. On examination I found the stricture low down almost cured, but as high up nearly as the finger could reach I felt a distinct contraction in the passage, and found a large part of the internal surface of the rectum wanting in the smooth feel of a sound mucous membrane. As there was still a slimy discharge, coming evidently from an unhealed sore, I directed an injection of nitrate of silver (gr. ij.—ʒj.) to be thrown up every other day. This was done two or three times, but unfortunately at this period the case took an unfavourable turn from a cause independent of the disease in the rectum. It appeared that since the formation of the abscess in the perineum, the stricture in the urethra had not gone on well, and that he had suffered from urinary irritation, passing his water in a small stream. He was very sensitive to pain, and refused to let the house-surgeon or dresser pass any instrument for him. He had been accustomed, before he came to the hospital, to introduce one for himself, and the dresser now allowed him to do so. After using one of small size he was attacked with acute peritonitis, which terminated fatally, November 2nd. On examination of the body there were found the usual marks of peritonitis, and an abscess in the areolar tissue of the pelvis, with a false passage from the urethra leading into it. There was a stricture at the bulbous portion of the urethra, the mucous membrane at the part being much torn. In the rectum there were two distinct ulcers and contractions. The upper ulcer was about three and a half inches in length, and seated at the extremity of the sigmoid flexure of the colon and commencement of the rectum. The lower and smaller one occupied the termination of the rectum: it was irregular in form, but narrow, varying in width from one to a half inch, and a narrow strip extended two inches up the passage. The contrast in colour between the ulcers and sound mucous membrane was very marked; the former presenting a dark slate colour, the latter a pinkish hue. The edges of both ulcers were in many places smooth and levelled off continuous with the floor of the ulcer, which was coated with distinct patches of condensed contracted indurated tissue, indicating partial repairs. In other parts the margin of the ulcer was irregular and undermined. The muscular fibres of the rectum were at parts bared and distinctly seen, especially in the lower ulcer. About eighteen inches of the colon above the upper stricture were examined and found healthy.

Though contractions in different parts of the large intestines have been noticed by pathologists who have investigated chronic dysentery in warm climates, this disease is by no means a common cause of stricture of the rectum.[5] Sir James Annesley and Sir Ranald

[5] In fifty-five cases of chronic dysentery examined after death in the General Hospital of Calcutta, Dr. Macpherson found the

Martin have borne testimony to the rarity of stricture in the rectum as a sequel of dysentery, and this view is confirmed by the observations made at the Dreadnought Hospital, where cases of chronic dysentery are of frequent occurrence.

Stricture of the rectum is a disease of middle life. I have already mentioned the case of a girl five years of age, and another fourteen, who had stricture in consequence of an injury, but it very seldom occurs in children. Bushe met with the disease in a lad nine years of age, and some years ago, a girl, aged eleven, died in the London Hospital from stricture and ulceration of the rectum, the history of which I have not been able to trace. Case 26 is an instance of a girl, aged thirteen, who had suffered from stricture in the rectum quite four years, consequently since the age of nine. This is the earliest age at which I have met with this complaint. It is rare, also, in old people. Most of the cases that have fallen under my notice have been between the ages of twenty and fifty. In the table of twenty-eight of my cases, already referred to, twenty-six were met with at this period. Two were older, one fifty-two, and the other seventy-six.

The earliest symptom of stricture is, generally, habitual constipation, with difficult defecation when the motions are solid. The difficulty being readily relieved by a solvent purgative, the nature of the case is not usually suspected at this early period. As the contraction increases, the constipation is with difficulty overcome, and the patient acquires the

colon contracted in three, and the cæcum nearly closed in one, but no mention is made of the rectum.

habit of straining to relieve the rectum. The stools are observed to be small in calibre, and are voided in small lumps.⁶ The mucous surface, irritated by the disturbance in the functions of the rectum, becomes inflamed and excoriated. This renders the actions of the bowels painful, a burning sensation lasting frequently for an hour or more after a stool. There is also a secretion of brown slimy mucus, which escapes with the motions, and soils the linen. The gases evolved in the intestines not escaping readily, give rise to flatulent distension of the abdomen especially in the course of the descending colon, and disagreeable efforts for relief. The bowels often remain constipated for days together, and then a spontaneous mucous diarrhœa, excited by the fæcal collection, or by a strong cathartic, softens the motions, and enables the patient to void the accumulated mass, its passage being attended with pain. In other instances, the patient is teased with frequent evacuations, fluid, and small in quantity. As the disease makes progress and ulceration ensues, the discharge becomes purulent and bloody, and the sufferings are much increased; the passage of motions being sometimes likened by the patient to a feeling as if boiling water were passing through the rectum. At this period, pain is often experienced

⁶ I give no account of the small, or flat tape-like, or figured fæces described by writers as characteristic of stricture, as I do not ascribe much importance to these appearances. When the bowels are irritable, and act frequently, persons with a healthy rectum will pass small and figured fæces; and an irritable sphincter likewise influences the size and shape of the motions. Besides, there is no necessity to pay much attention to an uncertain symptom, when an examination can so readily determine the real condition of the part.

in the sacrum. There is sometimes so copious a discharge as to mislead the practitioner, the stricture being overlooked, and the case treated as one of protracted diarrhœa. A slimy fluid, perhaps, escapes when the patient rises in the morning; and may, also, occur when he coughs or sneezes. The ulceration often leads to abscess and fistula, feculent matter being forced, or finding its way through the ulcer into the areolar tissue around, and exciting inflammation and suppuration. Fistula in ano, and sinuses in the buttocks or labia are, indeed, common complications of strictured rectum, especially in long-standing cases.

The appetite often remains good, and even the general health but little impaired, for a long time. The disease is very chronic in its progress; and so long as a passage for the motions can be obtained, though with difficulty, the patient continues following his avocations, suffering more or less at different periods. Indeed, it is surprising how great a length of time the general health will sometimes continue without being materially affected, even in cases of close contraction of the gut. The derangement of the digestive functions, the irritation kept up by the disease, and the exhausting discharges from the lower bowel, in the course of time, however, undermine the constitution, and bring on hectic symptoms. The appetite at length fails; there is sometimes urgent thirst; the body emaciates; night-sweats become profuse, and the stricture directly or indirectly becomes the cause of death. This is sometimes hastened by a lodgment of hardened fæces, or of some foreign body, just above the stricture, so as to block up the passage, and occasion all the ordinary symptoms of internal obstruction, with the death of the patient after many days' constipation. I know of several instances in which an occurrence of this kind first led to the detection of the complaint. In a patient whose motions are habitually soft, the stricture may make considerable progress without suspicion being excited of the existence of any important disease. He may continue for months subject to occasional constipation and derangement of the bowels, and passing fæces of small size, but experiencing no further inconvenience until a sudden stoppage and an examination of the rectum reveal the presence of a serious stricture. The suffering in stricture much depends upon the condition of the mucous membrane. When it becomes ulcerated early, there is generally more distress in the after-progress of the disease, and greater difficulty in conducting the treatment.

The symptoms of fully-formed stricture in the rectum are so clearly marked that the surgeon can generally predicate correctly the nature of the disease. It is necessary, however, to make a tactile examination. On exposing the anus, small flattened excrescences are usually observed at the margin of the aperture, especially when the stricture is seated near the outlet. These cutaneous growths resemble collapsed external piles, except that they are redder in colour, and are kept moist by the escape of a thin discharge from the bowel. They originate in the irritation kept up by this discharge. The finger, well greased, being passed carefully and gently into the rectum, will be arrested on reaching the stricture, so that the point only can enter. If the contraction be somewhat recent, and not very close, the surgeon may gradually dilate the **part, and, with** a gentle **boring motion of** the finger, penetrate **the stricture,** and thus examine its whole extent. **If he encounters** much **resistance,** or gives **much** pain, he must not venture to force **the barrier,** but must be content with ascertaining the seat and degree of contraction. **In** strictures high up in the gut, the rectum below may be found quite healthy, but it is often dilated and baggy with weakened expulsive powers.

In many cases of this disease the interior of the rectum is abundantly studded with small excrescences arising from partial hypertrophies or irregular growths of the surface and folds of the mucous membrane. **The** sensation communicated to the finger passed into the rectum is remarkable, the surgeon feeling **a number** of rough irregular eminences, more or less hard, thickly lining the surface. These excrescences, when numerous, have the effect of somewhat narrowing the canal **below the** stricture. This is situated further **from the** orifice **than** in ordinary **cases,** usually **at a** distance **of three inches. The** reddish flattened growths, resembling shrunk external piles, are almost constantly **found at the margin of the** anus in these cases. The changes in the mucous membrane above described are said to occur without any stricture. I have not myself met with any case of the kind. This disease is invariably attended **with** a profuse discharge **from the** rectum **of pus and slimy** matter mixed **with blood. There is not only painful tenesmus before a feculent evacuation, but a frequent and** urgent **desire to void** the slimy pus **and mucus** which collects **in** the bowel. This was so frequent and so pressing in **a** gentleman who was under my care, that he was unable to **go into** society, or ride in a public conveyance, or travel by rail.

Sir B. Brodie[7] and Mr. Colles of Dublin[8] regarded these growths occurring in stricture as constituting a peculiar form of the disease. Desault supposed that they were of syphilitic origin. A modified view of their syphilitic character has also been taken by M. Gosselin, in a paper,[9] founded on the observation of twelve cases, all females of loose life.[1] Without regarding the disease in the rectum in these cases as specific or as a remote consequence of syphilis, he considered the stricture and the growths from the mucous membrane to be the result of an extension of inflammation from primary disease near the anus to the rectum. In several cases of stricture combined with these excrescences on the mucous surface of the bowel, which have fallen under my notice, there has been no trace of constitutional syphilis, nor any other evidence of the rectal changes originating in specific disease. This condition of the rectum was well marked in Cases 26 and 27, which were clearly not of syphilitic origin. I believe, that this state of the rectum originates in chronic inflammation of the mucous membrane giving rise to a profuse secretion of a muco-purulent fluid at the lower part of the gut, a state somewhat analogous to chronic cystitis. The inflammation and discharge lead to the production of external piles, and excrescences and hypertrophies of the mucous mem-

[7] Lond. Med. Gazette, vol. xvi.
[8] Dublin Quarterly Journal of Med. Science, Feb. 1854.
[9] Archives Générales de Médecine, Dec. 1854.
[1] Mr. Colles gives a table of sixteen cases, and it is remarkable that thirteen were males.

brane, and higher up, by extension of inflammation to the submucous tissue, to the formation of stricture, the latter, probably, not being necessarily associated with the former. A communicated purulent discharge would be sufficient to give rise to these changes in the mucous membrane, and is probably sometimes the cause of them.

A stricture high up in the rectum is sometimes difficult of detection. In a case in which this disease is suspected, the bowel must be explored by a flexible instrument. An ordinary bougie passed into the rectum is commonly arrested on reaching the promontory of the sacrum, which might lead an inexperienced surgeon to conclude erroneously that contraction existed. When the passage is free, a good-sized flexible gum elastic tube may always be passed into the colon. The point is apt to impinge on the sacrum, or to be caught in a fold of the bowel, but if some warm fluid, water or linseed tea, be injected somewhat forcibly through the tube, a space is formed admitting the easy transit of the instrument. In stricture pain is felt when an instrument reaches the point of contraction, and a flexible one is arrested, or passed on with more or less difficulty. It sometimes happens, that the weight of the fæces accumulated above the stricture, especially if the patient is examined in the erect posture, or some violent straining, forces the contracted part low enough to be reached with the point of the finger introduced at the anus, the descent taking place in the form of a slight inversion of the bowel. A man with a stricture at this point was under my care in the London Hospital in 1850. The case was remarkable from the extraordinary dilatation which the bowel below the stricture had undergone. The finger seemed to pass into a capacious sac, at the fundus of which the contracted aperture in the intestine could be felt projecting.

In examining for stricture, the surgeon must recollect that the rectum is liable to be compressed and obstructed by disease of the neighbouring viscera,—by an enlarged or displaced uterus, fibrous tumours of this organ, an ovarian tumour, pelvic hæmatocele, an excessively hypertrophied prostate, or an hydatid tumour between the bladder and rectum. Cases of obstruction from all these causes have come under my notice in practice, and I have had occasion to tap an ovarian tumour and a pelvic hæmatocele through the rectum to remove the difficulty. I had under my care a female whose rectum was so encroached upon by a large tumour, apparently a fibrous growth from the uterus, that she was unable to pass any solid motion. Her bowels were never relieved until the fæces were rendered liquid by medicine. Several cases are recorded in which bougies have been long used for the cure of a supposed stricture in the rectum, when the obstruction has afterwards been found to arise from the pressure of tumours external to the coats of the bowel; and I have met with several cases in which a retroflected or retroverted uterus has been mistaken for a tumour in the walls of the rectum. The rectum may also be obstructed by an out-growth of fat, or by an infiltration of fat in the coats of the bowel. This is a very rare form of stricture. There is a specimen of it in the Museum of St. Thomas's Hospital, and Mr. Worthington has related a case of the kind in the Pathological Transactions. In the

Museum of the London Hospital also, there is a large fibrous and fatty tumour developed outside the rectum and contracting the passage.

Treatment.—The main object in the treatment of a stricture in the rectum is to remove the chronic induration, and to dilate the contracted part sufficiently for the free passage of the motions. The dilatation of the stricture is to be effected by mechanical means,—by the passage of bougies. Rectum bougies are made of a slightly conical shape, and of various materials; usually of **wax, elastic gum webbing,** or **caoutchouc. Wax bougies, being soft, are adapted for very sensitive strictures; but as** they **can seldom be used** more than once, and have little **effect on** a firm stricture, they **are not found so convenient** as the elastic gum, which is the best instrument for general **use.** Being smooth, **it glides readily through** the opening, and offers considerable **resistance to a firm** stricture. Before an instrument is used, the bowels must be well relieved either by medicine or an injection. The patient should lie on the left side, with the limbs bent on the body. The character **and closeness of the stricture being** ascertained by a careful tactile examination, a gum elastic bougie, of size sufficient to pass with ease, and without giving the least pain, should be selected. This, being warmed and well lubricated, should be passed in the most gentle manner through the stricture and fairly lodged within the sphincter. The bougie should then be retained for five, ten, or fifteen minutes, according to the character of the contraction. The operation must be repeated as soon as the irritation, produced by the instrument, has quite passed off, about every third or fourth day, and the size of the bougie may be increased according to the effect produced by the dilating process. This should always be very gradual, for forcible dilatation is very liable to excite inflammation in the coats of the rectum, and to aggravate the disease. Inflammation thus produced by a bougie has **been** known to extend **even to the peritoneum. The** treatment by dilatation **must be continued, not only** until an ordinary bougie of full size can be passed with **ease, and** the motions are evacuated of proper size, **but, even for** some **months afterwards,** at increasing **intervals in** order to counteract any disposition in the **contraction to return, and** to ensure, if possible, a **permanent restoration of the canal.**

An obstinate stricture in the rectum may be over**come** by forcible dilatation. For this purpose I used **to employ** bougies **made of** forcibly compressed dry sponge coated with tallow. When a small bougie of this kind is lodged in the stricture the tallow slowly melts away, and then the sponge getting saturated with moisture gradually swells, and gently and very effectively dilates the stricture. This treatment is open to the objection that in the removal **of the bougie, the swollen** portion of the **sponge in the dilated bowel above the stricture cannot be removed without being dragged with some force through the contraction, causing a** painful tearing sensation. In recent years, I have used in these obstinate cases bougies made of *laminaria digitata*, or sea tangle weed, which are very effectual in dilating the stricture, and can be removed without difficulty. A small bougie, with a ribbon attached for its withdrawal, is passed into the stricture and left **there.** It soon begins to swell **and** to dilate **the stricture,** and it may be taken away at the end of twelve or twenty-four hours. I usually order an opiate to allay uneasiness during the retention of the bougie. The regular passage of an ordinary bougie will be necessary afterwards. Several ingenious instruments have been contrived for the purpose of forcible dilatation, but they are all liable to more or less objection. In some there is great risk, in closing the blades, of the bulging mucous membrane being caught between them, and being torn in the withdrawal of the instrument; and in others the dilatation is unequal. There are also clever contrivances for dilating strictures by hydraulic pressure, and great power may be exerted in this way. In a case which I know of, the surgeon not estimating this power correctly, dilated until he burst the bowel, and the patient of course died of peritonitis. The rectum, when contracted and diseased, will not bear forcible and rough treatment, and inflammation and ulceration are readily set up by it, with severe aggravation of the symptoms of stricture. In cases of close, dense, organic stricture, which resists the action of bougies, dilatation may be facilitated by incisions, as in the cases to be narrated. Some surgeons recommend this operation to be done with the *bistouri caché;* but I prefer using a straight, probe-pointed bistoury, introduced flat upon the finger, and carried with it through the stricture. The blade can then be turned towards the contraction. An incision is usually directed to be made in the back of the rectum, towards the sacrum, but I find that more advantage is gained by three or four superficial notches in different parts of the contracted ring, than from a single deeper division of the stricture. To stop bleeding, and to keep the wounded structures apart, a plug of wet lint or cotton wool should be passed into the strictured part immediately after the operation, and retained there for a few hours; and gentle dilatation should be attempted on the next or following day. I have never met with hæmorrhage to any extent after the operation. It is very rarely that a vessel of any size runs directly beneath the mucous membrane in indurated stricture. Mr. Mayo, however, divided a stricture seated within three inches of the anus, towards the sacrum. The operation was followed in a few hours with very serious hæmorrhage, which was arrested by the introduction of a pledget of lint saturated with a strong styptic solution. A deep incision is not only liable to cause bleeding, which it may be difficult to stop, but also to lead to the formation of abscess and fistula, by allowing the passage of feculent matter into the areolar tissue about the rectum. Such an occurrence has happened several times after the operation, and, of course, has added to the difficulties of the case and distress of the patients. A case of stricture came under my observation, where a surgeon was induced to make an incision into it at the back part. The patient, a female, died about a week afterwards; and on examination I found a long sinus, containing a thin feculent fluid, extending from the wound upwards on the right side to the extent of six inches, and terminating under the peritoneum of the broad ligament of the uterus. There were marks of recent peritonitis in the pelvic cavity. In another case of close stricture, in a woman about forty years of age, where the incisions had been rather free, death followed from diffuse cellulitis and pelvic peritonitis. There was found a sinus leading from an opening in the rectum, no doubt made in the operation, into the areolar tissue at the back of the bowel. This tissue was infiltrated with lymph. If incisions be resorted to, they must be made with great care. Free and deep incisions are attended with very serious risk; and I know of one case in which, after two or three slight notches only, a large abscess formed behind the rectum, and burst into the bowel above the stricture. An exhausting purulent discharge continued for months afterwards. Of late years since employing laminaria bougies, I have nearly abandoned recourse to incisions.

In cases of stricture at the junction of the colon with the rectum, without any descent or prolapsus, the seat of contraction may be indicated by the limited distance to which a flexible tube can be passed, its reflection on reaching that point, and if the contraction be close, the return of injections. A long wax bougie, rendered flexible by warmth, and slightly twisted, may be insinuated through a stricture at this point, and I have succeeded in several cases in passing a very flexible gum elastic bougie with much advantage. Still, when the stricture is out of the reach of the finger, there is no way of ascertaining its character, no satisfactory guide for the selection of a proper-sized bougie, or for using it so as to dilate the contraction; no means, too, of determining positively whether the disease is simple stricture, or that form of disease—the carcinomatous —which is not likely to be benefited by mechanical interference, and in which the use of instruments is attended with risk of perforation. Such an accident has happened, indeed, without any disease at all, an instrument having been forced through the healthy coats of the intestine in the attempt to penetrate a supposed stricture. In the Museum of Guy's Hospital, there is a preparation of a colon in a perfectly sound state, perforated by a bougie at the distance of fourteen inches from the anus. It was taken from a gentleman who had long suffered from derangement of the digestive organs. This being at length attributed to stricture of the lower bowel, was treated by the passage of a bougie, which had been forced through the intestine into the peritoneum, and had destroyed the patient. The colon has even been perforated with O'Beirne's tube. I was present at the examination of the body of a man who had suffered from obstruction in the bowels. It appeared that a hard-handed practitioner, in giving an injection, had forced an elastic tube through the upper part of the rectum, and injected the abdomen with turpentine and castor oil. These cases would lead a prudent surgeon to be very careful in the introduction of instruments any distance along the gut, and especially cautious not to employ force to pass what he supposes may be a stricture. It should also be borne in mind, that the intestine, unless diseased, is not a very sensitive part, and will bear a good deal of pressure and rough usage without the production of pain. This will account for the injury which patients have been known to inflict on themselves in the passage of instruments into the rectum. Some years ago, a man, aged thirty-nine, was admitted into the London Hospital on account of a close stricture of the rectum. A bougie was passed two or three times, and, for convenience, left in charge of the patient. Being very anxious to make progress, he rashly **ventured to pass the** instrument himself. Shortly **afterwards he was** seized **with symptoms of** peritonitis, **and died the** following **day. On examination of the body I found the usual** appearances **of active peritonitis, and about an inch** and a half from the anus, a firm, indurated stricture of the rectum, an inch in length. Just above the stricture there was a perforation in the bowel half an inch in extent; and two inches above this, another rent, somewhat larger, through **which a portion** of intestine was protruding. This **case shows that it is not quite safe to** trust a patient **to** pass **a bougie for himself.**

In addition to these measures **for** dilatation of the stricture, means must be adopted **to relieve the irritability** of the part, and to ensure the regular passage of **soft** evacuations. A suppository of ten grains of soap **and opium** may be lodged in the bowel at bedtime, and, if the motions **are** costive, some confection **of senna with sulphur or castor oil, may be taken** in the morning, in doses just sufficient **to obtain** an **action of the bowels without** purging, **which** invariably adds **to the** patient's distress. Castor oil is of great service **in the** treatment of this disease. In small doses it softens the feculent masses, and lubricates the passage, without weakening the patient. The chief objection to its use with many persons is the nausea **to** which it gives rise. But if **the** patient **perseveres, the** stomach gets accustomed **to the remedy, which it tolerates as it does the codliver oil, so that we find patients with chronic disease of the rectum** continuing to swallow it daily for weeks and months without any feeling of nausea, or impairment of the appetite. The diet should be nutritious, and consist principally of animal food, so as to afford a small amount of excrementitious matter. Cod-liver oil is an excellent remedy in these cases. It nourishes the patient, and softens the feculent discharges, often rendering aperients unnecessary. It is no needless caution to advise patients to be careful to avoid swallowing plum-stones. The accumulation in the distended bowel above the stricture may be prevented by the occasional passage of an elastic tube through the contraction, and the injection of half a pint of tepid water, or soap and water. It may be necessary to repeat the injection two or three times a week. When much pain has been experienced after stools, and the discharge is considerable and slimy, or tinged with blood, I have found a good deal of relief derived from the application of a solution of nitrate of silver, in the proportion of five grains to the ounce of distilled water, to the diseased mucous surface included in the stricture. This can easily be made by means of a camel-hair brush passed through a small glass speculum open at the extremity, and introduced as far as the stricture. In a very bad case of strictured rectum under my care in hospital, in which the consolidation was too great, and the mucous membrane too much diseased to admit of my attempting dilatation, the motions passed with much less suffering after a few applications of the nitrate of silver solution in this way. Anointing the mucous surface with the mild citrine ointment, applied by means of a thick camel-hair brush passed through a speculum, has also a good effect in correcting this morbid state of the membrane. Smearing the bougie with ointments, as commonly recommended, is not of much service, as the ointment gets rubbed off in passing the sphincter, and does not reach the part affected; but by using grooved bougies, opiate or belladonna ointments may be carried within the strictured part. When stricture of the rectum is complicated with fistula, it is generally better to defer any operation for the latter disease till the contraction in the rectum is removed. I have, nevertheless, divided a fistula at the time of incising a stricture, as in Case 27.

The diseased mucous surface of the bowel above the stricture not only furnishes a copious discharge, which helps to exhaust the patient's powers, but it is sometimes the seat of profuse bleeding. In the autumn of 1852, I attended, with Dr. Hess, a young married lady, who, after suffering for some years from a stricture in the lower part of the rectum, was attacked with alarming hæmorrhage from the bowels, which continued for several days. The bleeding evidently came from above the stricture, and it was suspected to proceed from a spot in the descending colon, which was very tender on pressure externally. The hæmorrhage was effectually stopped by repeated cold alum injections, carefully administered with the long tube passed through the stricture.

The effect of a bougie introduced through a stricture of the rectum, as in stricture of other mucous canals, is at first to stretch, but afterwards to cause a gradual absorption and removal of the indurated tissue producing the contraction—the condensed areolar or fibrous submucous tissue. Such is the effect of pressure in curing strictures in the urethra, but it must be admitted that its influence is much less marked in stricture of the rectum. The surgeon is very rarely consulted in the early stage of the disease, when the complaint would be likely to yield readily to treatment. In Case 21, a contraction consequent upon the healing of an ulcer was overcome easily by dilatation resorted to at an early period. But when an **organic stricture** is fully established, it is **generally admitted to be** most difficult of remedy. Dupuytren stated emphatically, "Bougies give relief, but do not effect a cure." Dr. Bushe remarks, "Though I have ameliorated the condition of **many a** poor sufferer, I have never **been fortunate enough to** cure **a single case."** He adds, "**I** know of no patient who was able to **leave off the use of the bougie** without an increase or return of his complaint."[2] An excellent practical surgeon, Dr. Colles, of Dublin, states, "I feel confident that **a perfect cure** of the **organic stricture of the rectum has not been effected by any plan of** treatment hitherto employed." He adds, "I have paid great attention to the use of bougies, **and** yet I must candidly declare, that, hitherto, I have not been so fortunate as to have effected a permanent cure in a single instance; nor have I had the good fortune to meet with any patient whom I *knew* to have been **afflicted with** this disease, who **had been cured by another surgeon."**[3] These writers have undoubtedly taken too unfavourable a view of the results of treatment.[4]

Lib. cit. p. 287.

[3] Dublin Hospital Reports, vol. v. p. 142.

[4] At page 75 I have mentioned a case of recent stricture of the rectum, caused by the application of nitric acid for a prolapsus. A cure was effected by dilatation continued at intervals for eighteen months. A few years ago a lady came under my care on account of a close stricture just within the sphincter, consequent on the

The following cases show that a close contraction, long established, may be cured by careful and prolonged management :—

Traumatics tricture cured by dilatation.

Case 25.—A girl, aged fourteen, was sent me by Dr. Andrew Clark in December, 1873, on account of a close stricture in the rectum. She had been troubled with difficult and frequent defecation, coming on gradually for about four years. It commenced after an accident from falling backwards on the seat on the projecting brass button of a perambulator. Her mother stated that she could pass only very small motions, and that she was compelled to go to the closet every hour or half-hour. She also occasionally passed blood and mucus. On examination I found a firm circular stricture about an inch and a half from the anus, into which I could pass only the tip of my finger. Having had the bowels well cleared with castor oil, I inserted a small laminaria bougie. This was removed after twenty-four hours, when I was able to pass my finger easily through the stricture. Elastic gum bougies were passed regularly afterwards, and I was soon able to introduce a No. 7. She continued under treatment, having bougies passed at increasing intervals for about two years. In January, 1876, Mr. Cottew, her medical attendant, informed me that her motions were of natural size and form, but that she was still subject to a slight slimy discharge.

Old organic stricture of the rectum cured, by incisions and dilatation.

Case 26.—In May, 1860, I saw, in consultation with the late too free removal of parts in an operation for internal piles. The stricture had been incised five times by different surgeons, but without any benefit, as it returned as bad as ever. By gradual dilatation repeated during three years she got quite cured.

Dr. Brinton, a lady about twenty-four years of age, who had been troubled with an affection of the rectum since she was a child, indeed as long as she could remember. She was tall, and much accustomed to horse exercise. She had suffered from rheumatic fever, which had affected the heart, and had lately been troubled with rheumatism in the joints. She was subject to a free discharge from the bowels of a purulent fluid mixed with blood. It occurred irregularly, and was a source of great annoyance, as she had not the power of retaining it. The discharge was increased by certain articles of diet, such as fruit and uncooked vegetables. Her bowels seldom acted without medicine, and after a relief she constantly suffered pain. On tactile examination I found at the distance of an inch and three-quarters from the anus a close annular stricture, into which I could pass only the tip of my forefinger. The bowel below was tolerably healthy, and the sphincter acted properly. May 28th I introduced a straight probe-pointed bistoury along my finger, and notched the stricture in several places until I could pass my finger freely through it. I then detected above the contraction a quantity of excrescences from the mucous membrane. After the operation I passed daily an elastic gum bougie, commencing with No. 6, and increasing the size. After a week the bougie was passed on alternate days. In about ten days the soreness consequent on the operation passed off, and defecation became much easier, but there was no diminution in the discharge. After dilatation with the bougie I then threw up an injection of nitrate of silver (gr. iss—$\mathfrak{z}$j.) through a tube passed beyond the stricture. It caused an urgent desire for expulsion, and could not be retained. In about ten days this injection was discontinued, and a decoction of oak bark substituted for it. Dilatation and injections were continued with occasional intervals until September. At this time she was greatly benefited. Defecation was free and ceased to be painful. The discharge was much diminished in quantity, and could be retained. A No. 9 bougie passed readily, and the growths from the mucous membrane above the stricture were much smaller. I ceased to visit her from this date until January, 1861, when I was requested to see her in consequence of her suffering from pain in the rectum, and a pressing desire to relieve it. On examination I could find no contraction whatever in the bowel, but at the back, about the seat of the former stricture, there was a rough surface and indentation, resulting from a superficial abrasion, about the size of a sixpence, and pressure at this part caused pain. I introduced a glass speculum, and applied a solution of nitrate of silver to the sore. This application was repeated twice at intervals of a week, and fully succeeded in removing the pain and tenesmus, indeed she was much relieved by the first application. On examination at the end of three weeks I found the surface smooth. All irritation had ceased. The discharge continued to diminish, and was free from blood, and her general health was improved. In April, 1862, I had occasion to see this lady on account of another affection. She then stated that the only trouble in the bowel of which she had reason to complain, was a very slight discharge at times but she was able to eat fruit and salads without suffering. I considered the stricture permanently cured, no bougie having been passed for upwards of a year and a half. She died in December, 1863, of heart disease, without having required any further treatment for disease in the lower bowel.

Double stricture of the rectum cured, by incisions and dilatation.

Case 27.—S. S——, a girl, aged thirteen, from Hampshire, came under my care in the London Hospital, April, 1862. She had suffered from a discharge from the bowel, and from painful and difficult defecation for nearly four years. On examination I found two flattened excrescences, one at each side of the anus, and at the distance of an inch and a half from the aperture, a close, firm stricture in the rectum, capable of admitting only the point of the little finger. She had just left a hospital where she had been under treatment three months, and discharged incurable. Having had the bowel well emptied by injections through a long tube passed above the contraction, I introduced a No. 6 elastic gum bougie, and directed it to be retained a quarter of an hour, and to be introduced daily. This was continued **for three weeks**, when she lost appetite, and complained of uneasiness about the rectum. The left labium became tender and œdematous, and at one spot felt soft and boggy, indicating the presence of matter. May 8, chloroform having been given, the external excrescences were first excised. The labium was then punctured, and a quantity of pus discharged. Finding a sinus with an opening into the lower part of the rectum, I laid the fistula open with a bistoury, and then took the opportunity of notching the stricture in four places, until I could pass my finger freely through it. I then felt a number of eminences thickly studding the mucous surface above the stricture. After two days a No. 8 bougie was passed and retained for a few minutes, at first daily, and afterwards two or three times a week. The wound went on healing, but the girl did not mend in health, and complained of a hardness and swelling in the lower part of the abdomen on the right side. This ended in an abscess, which was opened, and about an ounce of healthy pus discharged. She afterwards improved in health, but defecation continued somewhat difficult, and there was still a free muco-purulent discharge. Some weeks subsequently a round worm was voided, after which her appetite improved, and she gained strength. In August a second stricture was detected at the junction of the sigmoid flexure and the rectum. A soft, flexible, elastic gum bougie, well warmed and oiled, was carefully passed through this every second day, and allowed to remain ten minutes. The size was gradually increased until a No. 9 passed with ease. The lower stricture is now quite cured, but the surface of the rectum is still uneven from irregular thickening of the mucous membrane, and there is also some discharge of pus, and occasionally a pressure downwards in defecation. She remained in the hospital under dilating treatment at prolonged intervals till the end of February, 1863, when she returned home apparently cured.

Her general health was quite restored, and there was every promise of her remaining permanently well.

We sometimes meet, especially in hospital practice, with old, inveterate, and neglected strictures, in which the disease is too far advanced to offer any prospect of being benefited by dilatation. In such cases, something may be done to mitigate the sufferings of this distressing complaint by measures which I have described, but in spite of all our care and palliative remedies, the disease will continue to make progress, wearing out the patient's strength, and ultimately proving fatal. Being strongly impressed with the smallness of the risk attending the operation of colotomy, and with the great relief derived from the operation in cases of painful cancer, I resolved some years ago when any case of non-malignant intractable stricture came under my care to suggest colotomy to the patient, with the view of obviating the sad misery consequent on the disease, and of prolonging life in a state of comparative comfort. This was done in the following case :—

Intractable stricture of the rectum successfully relieved by establishing an anus in the left loin.

Case 28.—M. A——, aged twenty-six, a farm-labourer from Essex, was admitted into the London Hospital, December 3rd, 1865. He was sallow and emaciated, and had suffered from disease in the rectum for more than twelve months, and complained of great difficulty, with pain and straining, in defecation. He was also troubled with a constant watery discharge, more or less tinged with blood. He had been many weeks in St. Bartholomew's Hospital, where he had undergone some operations. On the 8th I examined the bowel, and detected a close stricture about two inches from the anus. It was so tight and firm that I could introduce only the point of my finger into it. The bowel below the stricture was thickly studded with vegetations, and there were some flaps of integument round the anus. The examination caused great pain. Dilating treatment was attempted, but it caused considerable pain, and proved ineffectual; and, as he was getting gradually weaker and worn out by the disease, I proposed to him to undergo lumbar colotomy, and explained the nature and object of the operation. Having consulted his friends, and being satisfied that he was getting worse, he consented to the operation, which was performed on February 14th in the usual way, the colon being previously well distended with linseed tea, injected through a long tube passed beyond the stricture. He recovered favourably from the operation. Fæces escaped readily from the artificial opening, and there was a free discharge of a slimy mucous fluid per anum. This was checked by injections of tannic acid. About a month after the operation he passed some liquid feculous matter by the natural passage for two or three days, but afterwards all evacuations took place from the loin. Injections of a solution of chloride of zinc were substituted for tannic acid. The discharge diminished a good deal. He improved in health rather slowly, became timid about any surgical interference, and left the hospital at the end of May. I saw nothing more of him until February, 1867, when he visited me at the hospital. He was looking pale, weak, and out of condition. He suffered no pain in the rectum, and no feculent evacuations occurred from the natural passage, but there was a slight discharge occasionally. He refused to allow me to examine the rectum. In the loin there was a considerable prolapse of the bowel, which was easily reducible, and subsided spontaneously when he was recumbent. He wore a belt. His bowels acted once, and sometimes twice a day. He did no work, and was on parish allowance.

The operation in this case fully answered the end

in view in relieving the patient of the pain and other troubles resulting from the bad stricture in the rectum. His constitutional improvement was not so great as I had hoped for, but this must be attributed to an unhealthy constitution, and to his indigent circumstances preventing his procuring a sufficiency of good, nutritious food.

In August, 1865, I performed colotomy on an unhealthy woman in the London Hospital, on account of an intractable stricture and ulceration of the rectum with a fistulous opening into the vagina, at the patient's urgent request, for she had heard of the above case. I did the operation very reluctantly, as she had an enlarged liver. She took chloroform. The operation was followed by persistent vomiting, which lasted until her death on the eighth day from exhaustion. She had a large fatty liver, which weighed nine pounds and three-quarters. There were no marks of peritonitis.[5]

Since the publication of these cases colotomy has been performed for the relief of intractable and venereal strictures of the rectum by several surgeons in London, and by Dr. Mason in New York. The latter, in a valuable paper on colotomy,[6] gives in a Table twelve of these cases, including the above two, and three in which he was the operator. Of these eight recovered.

[5] See London Hospital Reports, vol. iv. 1867.
[6] The American Journal of the Medical Sciences, 1873, p. 354.

CHAPTER XIII.

CANCER OF THE RECTUM.

THE coats of the rectum are subject to carcinomatous degeneration in the three forms of *scirrhous, encephaloid,* and *colloid*. The scirrhous or fibrous form is sometimes developed in the submucous areolar tissue encircling the bowel at a particular spot, so as to lessen the area of the passage, and to produce an annular stricture. Either of these forms of cancer, may, however, invade the coats to a greater extent, contracting a considerable portion of the canal irregularly. Thus, in one instance of scirrhus which I examined after death, the rectum was diseased to the extent of two inches and a half. The upper opening would scarcely allow of the entrance of a small goose's quill; the lower would just admit the little finger; and between the two apertures the canal was irregularly dilated. Scirrhous degeneration may continue to increase until it narrows the gut to such an extent that only a common-sized probe will pass through it, and may at length completely close the canal. In the London Hospital Medical College, there is a fine specimen of colloid cancer, producing great thickening of the coats of the rectum, in some parts to the extent of an inch and a quarter, and stricture of the bowel. The mucous membrane within the contraction is the seat of a large ulcer. Fungoid growths sometimes spring from the mucous membrane at one side of the rectum, projecting into the bowel, yet interfering but little with the passage. The fibrous cancer and the soft medullary not seldom become blended together. Thus, in the later stages of the disease, a fungous growth may arise from a part previously contracted by scirrhous deposit in the submucous areolar tissue. The rectum occasionally becomes blocked up and occluded by fungous masses; or the changes which take place may have a contrary effect, degeneration and softening causing the coats to yield, and so increasing the calibre of the canal; or the impediment may be removed by sloughing of the softened growth, and detachments of portions of the morbid mass. A description of the progress of cancer of the rectum, and of the changes that occur in its advanced stage, is a description of the disorganization and invasion of all the tissues of the part, and of the organs in its immediate neighbourhood, in various degrees in different cases. In some instances, the carcinomatous bowel becomes wedged in the pelvis, agglutinated and fixed to the surrounding parts, forming one mass of disease. Frequently softening and ulceration cause fistulous communications with neighbouring parts—with the vagina in the female, and with the bladder or urethra in the male; or the peritoneum may become perforated, and an opening made into the abdominal cavity. When the passage is contracted, the intestine above the seat of disease becomes, as in simple stricture, dilated and hypertrophied.

Carcinoma may attack any part of the bowel, but generally affects the lower portion within three inches from the anus. It is liable to be developed also, though less frequently, at the point where the sigmoid flexure terminates in the rectum. The disease primarily developed in the intestine is frequently confined to this organ and to the adjoining structures, no other part of the body being found after death secondarily affected. But this is not always so. The lymphatic glands in the vicinity of the rectum often become enlarged; the liver is occasionally invaded by tubercles, and the peritoneum also studded with scirrhous deposits, and similar disease may be developed in the lumbar glands, and other internal parts.

Cancer of the rectum generally commences insidiously. Its early symptoms are, in many instances, so similar to those of simple stricture that the nature of the disease cannot be determined, or may not be suspected, until a considerable change has been effected in the condition of the bowel. The patient is troubled with flatulency, has difficulty in passing his motions, and strains in the effort to void them; and, as the disease makes progress, he experiences pains about the sacrum, which gradually increase in severity, and dart down the limbs. By this time some alarm is probably excited; and the surgeon, being consulted, will be led to make an examination. On introducing his finger into the rectum, he may find easily within reach a rigid contraction in the passage; but whether from cancer, or from chronic inflammatory thickening, it may be difficult to determine. The character of the pains may perhaps justify the more unfavourable conclusion. Should he feel any irregular nodules about the stricture, any hard solid tumour, or encounter a resistance like cartilage, or meet with softish tubercles which leave a bloody mark on the finger, then he would be able to pronounce on its carcinomatous nature. At a later period no difficulty is experienced. The surgeon feels a hard mass of disease, in which he may have some trouble in discovering the orifice of the passage, or finds rounded fungoid growths which bleed readily when touched. The disease may extend as low as the anus. An irregular red-looking growth sometimes protrudes externally, blocking up the passage or displacing the anus. The stools become relaxed and frequent, and contain blood, and, in passing, cause a scalding pain, and give rise to severe suffering. Often, also, there is a thin, offensive, sanious discharge. If the disease be largely developed internally, increasing difficulty may be experienced in evacuating the bowels; or, in consequence of softening having caused the parts to yield, it may be the reverse, the motions passing with less trouble. Loss of retentive power may ensue, adding greatly to the patient's troubles. This may arise from the disease invading the anus and destroying the sphincter, but loss of retention occurs also when cancer is developed higher up in the bowel, the lower part being free. This may be explained by the carcinomatous tumour pressing on or destroying the nerves supplying the sphincter, and so paralyzing the muscle. The sufferings increase; severe shooting pains are referred to the groins, back, or upper part of the sacrum, and often extend down the thighs and legs, leaving a dull fixed uneasiness in the intervals. The constitution suffers in due course; the patient exhibits the blanched sallow look, anxious countenance, and emaciated appearance, commonly observed in persons suffering from malignant disease. If complete obstruction of the bowel do not occur to accelerate a fatal determination, as not unfrequently happens, fresh troubles arise. In consequence of a communication becoming established between the rectum and urethra or bladder in males, flatus escapes from the urethra, and liquid fæces pass with the urine; and in females, motions are discharged at the vagina. The passage of part of the contents of the bowels **by these** unnatural channels greatly increases the misery **of** the patient's condition, rendering him **an object of** disgust to himself, and offensive to those about him. **An** ulcerated opening **into the** peritoneum, allowing the escape of feculent matter into the abdomen, may excite peritonitis, and thus bring **the** case to a fatal termination; **or, the** powers of **life** gradually giving way, the patient becomes hectic **and** exhausted, worn out by this painful and **distressing** malady. There is great variety, **however,** in the degree of suffering, and even of constitutional derangement, attending this disease. The sufferings are in some instances excruciating; in others, comparatively slight. I had a man under my care whose anus was blocked up with carcinomatous fungus, and who had an opening into his urethra; but the pains were not severe, nor had his constitution suffered to any great extent. His chief complaint was of gas escaping from the urethra. In a case in which I performed colotomy there was no constipation, and **no suspicion of any disease in** the rectum until **the sudden occurrence of obstruction, when a mass** of carcinoma was detected in the bowel. Cancer of the rectum occurs generally in middle life. The earliest age at which I have met with it is twenty-three. The patient was a young man in the London Hospital. The disease extended a considerable distance up the rectum, and formed a large characteristic cancerous **sore** entirely around the anus, but he suffered very little from it.

A woman with **cancer of the rectum may** become pregnant. **Some** years ago I saw, in consultation **with the late Mr.** Ward of Epsom, **a lady about** thirty-five years **of age, who was in** the seventh month of pregnancy, and had extensive cancerous **disease of the** lower bowel. As the pressure upon the parts during delivery at the full period was pretty sure to aggravate the disease, I had no hesitation in recommending the induction of premature labour.

It is commonly believed that cancer of the rectum attacks women more frequently than men. This does not accord **with my experience of cases seen in** hospital and private practice. **Of** twenty-one of the **latter** of which **I have** taken notes, seventeen were **males,** and four females.[7]

Little can be obtained from remedies in this terrible disease beyond palliation of **the** symptoms and **ease from pain. I have in** some few cases of cancerous stricture with partial obstruction, rendered **the passage more free by cautious** dilatation with bougies. Mechanical interference, however, is not free from **risk, unless great care be used. I have** met with one **case, and seen the** preparation of another, in which the practitioner, in using this instrument, passed it through the softened tissues into the abdomen, and thereby accelerated the patient's death by causing peritonitis. Some of the **German** surgeons have managed to free the passage **by scraping the cancerous surface with a sort of**

---

[1] Mr. Carter, formerly house-surgeon at St. Mark's Hospital, informed me that of thirty-five cases admitted during four years, nineteen were men, and sixteen women. Of eleven cases noted by Mr. Baker (Med. Chir. Trans. vol. **xlv.**), eight were males, and three females.

sharp spoon, any hæmorrhage which may ensue being checked by cauterization. Esmarch takes a favourable view of this treatment. A patient with cancer of the rectum should live a quiet life, and remain a good deal in the recumbent posture, and take a nourishing, but not stimulating diet. The general health may be supported by tonics. The bowels must be kept open, and the motions rendered soft, if necessary, by small doses of castor oil. If the stricture should be very close, so as to cause a lodgment of the fæces above, it may be necessary to pass a long tube through the contraction, and to inject warm water, soap and water, or warm olive oil, in order to break up the feculent masses. The greatest care must be used in the passage of the tube. In a hospital case of cancerous stricture, rather high up, in which I directed it to be employed as occasion required, the dresser, on the third or fourth time of using it, unfortunately passed the tube through the soft carcinomatous mass, and penetrated the abdomen, causing the patient's death in twelve hours. Pain can be alleviated by opiate and belladonna injections, or by small doses of chloral or morphia taken night and morning, their strength being gradually increased as the effects of the remedy diminish. Subcutaneous injections of morphia also give great relief.

The question naturally arises, can anything be done by operation in cancer of the rectum as in cancer of other parts of the body, to arrest the disease, or to protract life and relieve pain? I am indebted to Mr. Holt for the brief particulars of the case of a gentleman about fifty years of age, from whom he removed with the ecraseur, in 1874, a projecting **fungoid growth in** the rectum two inches above the anus. He divided the sphincter, dragged down with a pair of forceps the diseased wall of the rectum, transfixed the base beyond with a strong straight needle, passed the écraseur over it and removed the parts. There was no bleeding of any moment, the part healed over, and now after two years the patient is quite well. The growth was not subjected to minute examination, but Mr. Holt believes that it was cancerous, and Sir James Paget, who saw the patient before the operation, **was of the same opinion.**

After a long experience I must say that no case of internal cancer of the rectum has come under my notice in which it was possible to perform a similar operation. Patients rarely apply for assistance until the disease has extended too widely and too deeply to admit of extirpation with the écraseur or galvanic cautery. If such a case occurred to me, I should not hesitate to resort to one or the other mode of removal. Excision of the carcinomatous rectum was practised formerly by Lisfranc and Dieffenbach, and is resorted to in the present day by several German surgeons. Parts of the vagina, urethra, and bladder have been cut away in these operations, and it is related that life has been prolonged by them. We are told that incontinency is not an inevitable result, even if the anus be removed, and Wutzer is reported to have excised four inches of bowel without incontinency ensuing.[8] To credit

---

[8] Esmarch. Die Krankheiten des **Mastdarmes** und des Afters. Handbuch **des** allgemeinen und speciellen Chirurgie. Band iii. Abth. 2. Erlangen, 1872.

this we must suppose that the sphincter is an unnecessary muscle.

I am unwilling to discourage any attempt to cure or relieve so dire a disease as cancer of the rectum by excision, but knowing the danger which must be incurred from hæmorrhage in the operation, the misery likely to ensue from incontinency of fæces, and from contraction in the wound, if healing takes place, as well as the prospect of an early return of the disease, I cannot think that the chance even of a prolongation of life is worth acceptance on the terms offered by such an operation,[9] and I hold that much more is to be gained with less risk to life by another proceeding, viz. lumbar colotomy, which is applicable also to cases in which excision would be impossible.

In the third edition of this work (1863) I endeavoured to show the advantages which may result in cancer of the rectum from diverting the channel for the passage of the fæces by establishing an anus in the left loin, and in some observations published in 1865,[1] I argued that by this operation we should be able to prevent or mitigate many of the distressing symptoms of this terrible disease, and sometimes to retard its progress. I adduced three cases in which lumbar colotomy had been performed chiefly with this view. In one, a lady with a recto-vaginal fistula from carcinomatous ulceration, became almost entirely free from pain after the operation, and improved in spirits, complexion, and general

---

[9] I do not allude here to cases of cancer of the anus extending into the rectum. Many such cases are very suitable for excision, as I shall show in the next chapter.

[1] *Lancet*, Jan., 1865.

condition, and was able to go out daily in a bath chair. She survived the operation three months. The second case was a man aged forty-five, who suffered considerably in the back and loins, particularly after an evacuation, which was attended with agonizing pain. The symptoms were much relieved by the diversion of the fæcal matter through the artificial anus. He survived eight months, and died exhausted. The third case, a man aged twenty-nine, came up to the London Hospital from Yorkshire, suffering severely from cancer of the rectum, which had commenced sixteen months previously. It had advanced until he had lost all power of retaining his fæces, and he was subject to a constant flow of feculent matter with mucus, and sometimes blood. He recovered favourably from the operation. The artificial anus answered well, his motions being regular, the bowels acting twice daily. He soon lost the acute pain in the back, and five months later his appearance was healthy, and he had gained a stone in weight. He subsequently lost flesh and strength, and died from exhaustion nine months after the operation. Since the publication of these cases colotomy has been performed in painful cancer of the rectum without obstruction, by myself, in six additional cases, and by many other surgeons, with the result of prolonging life far beyond the periods above mentioned, so that colotomy may now be regarded as an established operation for the relief of this distressing disease.[2]

[2] Amussat long ago suggested the operation for delaying the progress of cancer of the rectum, and prolonging life, but it does not appear that he ever performed it.

CHAPTER XIV.

EPITHELIAL CANCER OF THE ANUS AND RECTUM.

THE anus is liable to a form of cancer, the *epithelial*, which is apt to attack those parts of the body where a junction takes place between the skin and mucous membrane, as the lips, the eyelids, the prepuce, and the extremity of the penis. The external characters of epithelial cancer at the anus are the same as those observed in other situations, and the disease usually extends into the rectum. The following are well-marked examples of this affection, and indicate the best mode of dealing with it.

Epithelial cancer of the anus and rectum cured by excision, after repeated failures of treatment.

Case 29.—Mrs. M——, aged forty, an English lady, married to a German professor, but without having borne children, consulted me in April, 1855. On examination I found a large, elevated, and slightly indurated sore, occupying the whole of the right side and part of the back part of the rectum, just within the sphincter muscle, and extending up the bowel the distance of about an inch and a half. The sore was somewhat larger than a crown piece. There was slight bleeding from the surface after the removal of the finger. The chief symptom she complained of was a frequent smarting pain which became more severe after an evacuation. At this time there was usually a slight discharge of blood. There was no obstruction in the passage. The lady looked pale and anxious, but in other respects seemed free from disease. It appeared that the complaint of the rectum was first noticed two years before. At that time she was residing in Germany, and she consulted the late Professor Siebold, of Jena (Saxe-Weimar), who excised the diseased part in September, 1853, whilst she was under the influence of chloroform. She recovered slowly from the operation, and remained apparently well until July, 1854, when a return of the disease was noticed, and the complaint shortly became as painful as before. She subsequently went to Paris, and in August placed herself under the care of a German surgeon practising there. He made repeated applications of a caustic nature to the sore, and finding them unsuccessful, at length proposed the actual cautery, which was used in February, 1855, the patient being under the influence of chloroform. She remained under the care of this gentleman altogether six months, but, according to her account, she derived no benefit from his treatment, and was not free from the shooting pains any part of the time. She was induced, therefore, to come to London for further advice, and, at the recommendation of Dr. Swayne, of Clifton, consulted me. I entertained no doubt respecting the nature of the disease, and proposed the operation of excision, but considering the failure of the treatment previously adopted, I advised her husband to take another opinion. Mr. Hilton, a few days afterwards, met me in consultation, and fully agreed with me that the disease was epithelial cancer, and could be entirely removed with the knife. The disappointment which they had experienced naturally led both the patient and her husband to distrust a repetition of excision. I consequently saw nothing more of them for a month, during which period they sought other advice, and also communicated my proposal to Professor Siebold, who wrote and recommended her to submit to the operation, when they again applied to me. On May 30th, 1855, the bowels having been well relieved, the patient took chloroform, and I then excised the growth, taking care to cut wide of the disease. In doing so, I removed nearly the whole of the sphincter muscle on the right side. By carrying the point of the forefinger of the left hand beyond the upper margin of the ulcer, and cutting over it, I made sure of excising completely the portion of the disease which was deeply seated in the rectum. Several large arteries which bled freely were at once secured. This was attended with some little difficulty, owing to their depth in the pelvis consequent on the retraction of the **levator** ani muscle. The wound was afterwards carefully plugged. **No** unfavourable symptom followed. The wound **healed** very slowly, but steadily; and, by August 9th, **had quite closed.** For some weeks after the operation the patient **lost the** power of retaining the fæces, **but it** was regained by **the** time the wound closed, except **when the** bowels were much relaxed. **The contraction at the anus** was much less than might be **expected, considering the amount of substance, and** of **the sphincter muscle, removed. The aperture admitted the** passage **of the forefinger without** difficulty. **Seven** years after the operation, I had **the opportunity of ascertaining** that there had been **no return of disease in the part,** but during a twelvemonth she had been a sufferer from **a** tumour of doubtful nature high **in the pelvis.** The diseased part, when examined in the microscope, exhibited the characters **of epithelial cancer.**

Epithelial cancer of the anus and rectum removed by excision.

Case **30.**—E. C——, a **stout, married** woman, aged **forty-nine,** the mother of several children, of pale complexion, but in tolerable health, was admitted into the London Hospital, January 11th, 1855. She had suffered from what she believed to be piles for about sixteen years, and had been subject to bleedings. About three months before her admission her surgeon excised a tumour from the anus, **which** she described as being the size of a hen's egg. **The part** healed, but afterwards ulcerated, giving rise to the present disease. Since its formation she had suffered sharp irregular pains in the part, and soreness during the passage of stools. None of **her** family had suffered from **cancer.** On examination, I found an ulcerated sore occupying the right side of the anus, and extending some distance into the rectum. It was about the size of a crown piece, and not very hard. Its **edges** were raised, ragged, and slightly overlapping; its **surface irregular.** A small piece detached from the sore, **and** examined in the **microscope,** exhibited the characters of **epithelial cancer.** There were also some warty growths in the vicinity of the large sore, and on the opposite side of the bowel, but they were neither hard nor ulcerated, and I did not regard them as cancerous. On the day after her admission, the bowels having been well relieved, I excised the cancerous growth, taking away a considerable portion of the sphincter muscle on the right side. There was smart hæmorrhage from several vessels, their orifices being retracted and deeply seated. With some trouble they were secured, and the wound was afterwards plugged. She bore the operation very well without chloroform, which she objected to take. An astringent draught with opium was given after the operation, the bowels remaining unrelieved until the fifth day after the operation, when they were acted on by castor oil. She quite lost the sharp pains, and the wound soon began to heal. The soft warts about the anus were touched with the potassa fusa, under which application, repeated three or four times, they gradually disappeared. She was discharged from the hospital on the 14th of April, the wound being quite healed. The anus was contracted, but it readily allowed the passage of the forefinger, and no difficulty was experienced in defecation. She was also able to retain her motions as before. There was still a strong disposition to warty growths about the anus; and after her discharge from the hospital she returned occasionally to have the potassa fusa applied to them. A lotion of the nitrate of silver was also kept to the part. After a time she ceased to attend, and in January, 1856, she was again admitted in consequence of a mass of soft warts having sprung up close to the cicatrix at the anus. They were not ulcerated, and caused no pain, but being apprehensive that they might undergo cancerous degeneration, I thought it desirable to remove them, and they were excised on the 28th, under chloroform. They proved to be simple epithelial growths, or hypertrophy of the normal elements of the part. The wound healed favourably, and she was discharged in the beginning of February, and recommended to keep a rather strong nitrate of silver lotion to the part. The tendency to warty productions in the skin above the anus, though partially restrained by the lotion, was, however, quite remarkable, and in September of the same year she was admitted into the hospital a third time, on account of fresh growths having arisen. They were slightly prominent, and exactly similar in character to those removed in January, and free from ulceration. The warts were removed this time by the repeated application of a caustic composed of muriate of antimony, one part; chloride of zinc, one part; and plaster of Paris, three parts. This composition formed a sort of paste very convenient for use, but it caused a good deal of pain, which lasted some hours, and had to be alleviated by full doses of opium. She remained in the hospital until the middle of November. The warts had not entirely disappeared, but she was anxious, on account of her family, to return home. In February, 1857, I again admitted her on account of large flattened warty growths around the anus, in two considerable masses, and one small one. There was ulceration on the surface of one of the former, with some amount of induration, and this was the seat of a good deal of pain. On the 12th I touched one spot, near the verge of the anus, with some strong nitric acid. In a few days the nitrate of silver lotion was applied to the wound, which healed favourably, without further contraction of the orifice, and all pain ceased. There was afterwards some indication of a rising of fresh warty growths, but it was checked by the application of strong nitric acid. I should now have discharged my patient cured, but for some weeks a glandular swelling had been forming in the neck, on the left side, just beneath the lower jaw, and it ended in an abscess, which was opened on the 26th of March. About a week afterwards she was seized with erysipelas of the face, which unfortunately had a fatal termination on the 8th of April. The body was examined, but there were no enlarged glands, indeed no internal organic disease.

This case is remarkable for the strong tendency to the formation of warty growths, a tendency limited to the skin in the immediate vicinity of the contracted anus. The mass removed in January, 1856, a year after the excision of the epithelial cancer, consisted of areolar tissue, hypertrophied papillæ, and enormously accumulated epithelial cells of the cutis. From the fact that these elements had preserved their normal relations; that the cells had not invaded the subjacent tissues; that there were no nest, or granule cells, and no heterogeneous forms of any kind, it was inferred that the growth was innocent. The warts which sprang up afterwards were removed by escharotics, which caused considerable pain, and my experience of the action of caustics on morbid growths connected with the skin, convinces me that they produce, as in this instance; more suffering than the knife. It was very necessary to get rid of these warty growths as they were renewed, not only on account of the irritation they produced, but also because of their liability to degenerate into cancerous disease.

These two cases show, that a large part of the sphincter muscle may be excised without seriously weakening the retentive power of the anus, or contracting the orifice so as to produce any important impediment to the passage of stools. Excision is the treatment best adapted to the entire removal of an epithelial cancerous growth of any great size in this part. Powerful caustics, even the actual cautery, failed to obtain a cure in the case of Mrs. M. There is this great advantage in the knife, that the surgeon can make pretty sure of thoroughly removing all existing disease; whereas the extent of the operation of a caustic is somewhat uncertain: it may destroy too much or too little. It may be objected, that in Mrs. M.'s case, the first operation of excision was not successful, the disease having returned; but it seems highly probable that Professor Siebold was not then aware of the real nature of the lesion, and regarding it as an innocent growth, was not so careful to excise freely all the morbid parts. We have some ground for this conclusion, not only from the rarity of the disease, but also from the circumstance that this distinguished surgeon, when appealed to twenty months afterwards for his opinion respecting a repetition of the operation which I then proposed, advised its performance. The length of time, seven years in the first case, and upwards of two years in the second, which elapsed after the operations, without a recurrence of the cancerous disease, is sufficient to show that in each instance the growth was entirely removed.

In excising growths at the anus extending into the rectum the surgeon cannot be too careful in guarding against the dangers of hæmorrhage. If the excision reach above the sphincter, the parts are liable to be deeply retracted by the levator ani muscle, and blood may be largely poured out into the bowel without exciting alarm. A woman, aged thirty-one, had a large epithelial cancer excised from the anus by an able provincial surgeon. There was profuse hæmorrhage at the time, but it was quickly checked by the use of the actual cautery. She never rallied well, and death occurred on the fourteenth day, from diffuse suppuration and gangrene.[3] In August, 1860, I removed a large epithelial growth extending into the rectum from a gentleman, aged sixty-six, who also laboured under disease of the

[3] Medical Times, vol. i. 1860, p. 498.

heart. There was not much bleeding at the time of the operation, but I was careful to leave an assistant in charge of the patient. In two hours I was summoned in haste, in consequence of some large evacuations of blood by stool. The bleeding came from a large artery at the upper end of the wound, which was retracted so deeply that the surgeon in charge could not manage to secure the vessel without assistance. I felt it pulsating, and with a pair of dressing forceps seized the part, dragged it down to the anus, and my assistant applied a ligature. No further bleeding ensued, and the patient went on well for a week, when he was attacked with violent diarrhœa, and during a copious evacuation, two days afterwards, he expired suddenly. In cases of hæmorrhage, after the operation, if the surgeon cannot find the bleeding vessel, or manage to secure it, he must plug the wound in the way directed at page 100. I have found this quite effectual, and have never had occasion to employ the actual cautery.

Though epithelial cancer usually commences at the anus, it may occur in the mucous membrane of the bowel. Some years ago, I met with a case of an old woman, in which this disease attacked a portion of the ileum some weeks after its implication in a strangulated hernia relieved by operation. The following is a case of

Large epithelial cancerous ulcer in the rectum.

Case 31.—Mrs. R——, a stout lady, about sixty years of age, from Liverpool, consulted me in February, 1856, supposing that she was suffering from piles. She looked in pretty good health, but complained of occasional pains in the rectum, especially after a solid motion, although there was no impediment to defecation. The pains were not constant nor severe, and did not disturb her rest at night. She was also subject to slight **discharges of** blood. On examination, I discovered, just within **the sphincter,** a large elevated sore on **the left side and back of the rectum.** It was not very hard nor tender. I could just reach its upper extremity with the point of my forefinger. As well as I could gather from the patient, the disease had existed about two years. She was prepared to submit to any treatment I might recommend. But as the complete excision of so considerable a growth from the rectum would not be free from risk, especially in a person somewhat advanced in life, and as the disease was making but slow progress, caused but little suffering, **and during two years had scarcely affected her constitution, I considered that her** welfare would be best consulted **by the use of means calculated to maintain her general** health and to palliate her **symptoms.** I prescribed some **tonic medicine with mild laxatives, and applied to the sore, every other day,** a solution of **the nitrate of silver. This application** had the effect of relieving **the sensation of soreness** and of diminishing the bloody discharge. She also took **small doses of** morphia when the pains were **greater** than usual. **After** a week, a growth from the lower border of the ulcer projected at the anus, and caused her pain in sitting and at other times. As this was likely to increase and give more trouble, I excised it with a pair of curved scissors. **The** cut surface bled somewhat freely, and it was necessary to tie two vessels. The operation relieved her a good deal; bleeding from the sore ceased; and after remaining under my care a few weeks, she returned home. The characters of epithelial cancer were clearly seen on the examination of the portion removed. I have not seen her since; but in February, 1858, two years later, her surgeon, Mr. Edgar, in answer to **my** inquiries respecting **her** condition, informed me that the disease entirely surrounded the anus, and extended about two inches **up the** rectum. It produced but little impediment to or pain in defecation; and at other times her only suffering was from a soreness, which was aggravated by exercise. Slight hæmorrhage has occurred **several times,** and on two occasions it was considerable. The last time it took place to the extent of a pint, and was with difficulty restrained by means of styptics and a compress. There was no enlargement of the inguinal glands, nor evidence of internal cancerous disease. Her general health had not suffered much. Mr. Edgar added that she appeared as well as when she was under my care in London. Her appetite was good, and she slept well. She died in the following November, about five years after the disease was first noticed.

Epithelial cancer of the rectum differs in many respects from the malignant diseases more commonly affecting this part. Scirrhous and medullary cancer generally produce, sooner or later, some contraction or obstruction in the passage, and show a tendency to involve the parts around. In cases of epithelial cancer I have seldom noticed any impediment in defecation, and have generally found the passage free and unobstructed. Neither do patients complain of the distressing pain, referred usually to the sacrum, which persons affected with scirrhus of the rectum so commonly experience, nor of the painful tenesmus and defecation which add so much to their sufferings in this form of the disease. There is also an absence of the cancerous cachexia, of the emaciation and pale and anxious countenance so frequently remarked in malignant disease.

An early recognition of an epithelial cancer at the anus and lower part of the rectum is very important, since it is a disease which may be effectually removed by operation. When the surgeon meets with a raised ulcer, with an uneven surface and indurated edges, causing, if it extend into the rectum, but little pain, and producing no contraction in the passage, he may suspect that the disease is an epithelial cancer, and treat it accordingly. Its true character can only be determined with accuracy by a microscopical examination of the morbid tissue.

CHAPTER XV.

MELANOTIC CANCER OF THE ANUS.

No case of melanotic cancer at the verge of the anus has fallen under my notice, though this is a part which we might expect would be liable to the disease. A case of the kind came under the care of Mr. Moore at the Middlesex Hospital. A man, aged sixty-four, had a fungating melanotic growth on the right side of the anus. It was ulcerated and bled frequently. The disease, including three-fifths of the external sphincter, was excised. The part healed soundly without loss of control over the bowel. The patient remained well for a year, when disease appeared in the rectum, and produced cachexia, and the usual symptoms of malignant disease of the bowel, but there was no return at the seat of the operation.'

As in epithelial cancer, an early excision of the morbid parts is the treatment required.

CHAPTER XVI.

OBSTRUCTIONS OF THE RECTUM, AND OPERATIONS REQUIRED FOR THEIR RELIEF.

I HAVE stated, that a stricture sometimes forms

' Medical Times, March, 1857.

high up in the rectum, at the point where the colon joins it, and that this part is also subject to cancer. Diseases in the bowel so far up as to be beyond the reach of the finger are almost sure to escape detection until the occurrence of serious symptoms. Indeed, the inconvenience resulting from stricture in this situation is often so slight, that no suspicion of its existence is excited until the bowel becomes obstructed by the impaction, at the contracted part, of hardened fæces, or some solid body, as a plum-stone. When obstruction occurs from a stricture near the anus, the surgeon is able to extract or dislodge any substance so blocking up the orifice, or, in case of extreme contraction, to afford relief by dilatation or incision, and injections through a flexible tube. But if the impediment exist, as it more frequently does, at the termination of the sigmoid flexure, and out of reach of the finger, he will probably fail in his attempts to remove it, and the patient's life then becomes exposed to imminent danger from insuperable constipation.

In the early stage of obstruction, whilst doubt exists respecting its nature and seat, purgatives and injections are usually given to overcome constipation. These means, having been fully and fairly tried, should not be continued. The return of the injections generally indicates the probable seat of the difficulty. In obstructions of the large intestines, if the patient be judiciously treated, and if his stomach has not been upset by repeated doses of drastic purgatives, a long period may elapse before the vital powers give way. Patients have been able to take fluid nutriment in small quantities, by the mouth, such as wine, beef tea, and milk, and have been nourished by injections of the same, and thus supported have lived four, five, and six weeks, and even three months without passing any stools. During this interval, the propriety of having recourse to an operation to provide an artificial vent for the fæces must necessarily come under the consideration of the surgeon. The first point to be cleared up is, all doubt in respect to the seat of obstruction. It may be found, that only a small quantity of fluid can be thrown into the bowel, and that it readily returns uncoloured; that the long flexible tube will not pass further than about eight inches;[5] or, if its progress be not arrested at that distance, that the finger introduced into the rectum, by the side of it, will meet the end of the tube, which, on reaching the obstruction, has turned back. The distended colon may be traced down into the left iliac region. These signs, especially if accompanied with pain referred to, or felt on pressure at the upper part of the sacrum, towards the left side, would pretty clearly indicate the exact situation of the obstruction. And, as an impediment very seldom occurs at the point of termination of the colon in the rectum from any other cause than simple or malignant stricture,[6] the surgeon becomes nearly apprised of the nature of the

[5] In exploring the large intestines in cases of obstruction, the surgeon should use a long flexible tube, not too large, with an aperture at its extremity. The extremity is liable to be arrested in its course by catching in the folds of the bowel, and by coming in contact with the promontory of the sacrum. If a syringe be fitted to the tube, and an assistant pump in some warm linseed tea or water as the tube is passed on, any accidental impediment of this kind will be avoided or removed.

[6] Intussusception occurs at this part; but in such a case the invaginated intestine would be felt in the rectum. Accumulations

case with which he has to deal. His opinion will be strengthened, if he finds, upon inquiry, that the illness has been preceded by slight attacks of constipation, and difficulty in regulating the bowels. It is right to add, that notwithstanding these guides, the diagnosis may be difficult. In a case of stricture at the termination of the colon in the rectum, which came under my notice, some of the surgeons consulted hesitated pronouncing a positive opinion as to its seat. In another case of internal obstruction which was operated on without success, the surgeons were completely mistaken; the distended small intestines occupying the pelvis having so pressed on the rectum as to prevent the lodgment of injections, and to cause the doubling of the long tube, which led to the supposition that the obstruction was at the extremity of the colon, instead of in the ileum, as appeared after death. In a dilated or lax state of the sphincter muscle, a small hand well greased may be slowly insinuated into the rectum so as to reach and determine the nature and exact seat of obstruction.

Cases of obstruction requiring operation occur more frequently from cancerous disease than from simple stricture. Any part of the lower bowel is, indeed, liable to become so contracted and blocked up by carcinomatus growths as to prevent the passage of fæces.

The knowledge of the cause of the obstruction, and of its seat in the lower part of the alimentary canal, places these cases in a different category from those of internal obstruction, in which, with the

---

of hardened fæces above the same point would be dislodged by introducing the long tube, and throwing up injections.

utmost skill and care, and under the **most favourable circumstances, the diagnosis of** the situation and nature of the impediment must always be involved in considerable obscurity. Besides, there is not the same occasion for delay in the hope or chance of the impediment yielding, which tends so much to embarrass the **practitioner** in treating the more doubtful cases; **for,** when the ordinary means of giving relief have failed, it is **clearly** the duty of the surgeon to suggest the expediency of the operation **for an** artificial anus before inflammation is set up, **or the intestines** have become damaged by over-distension, **or** before the powers of life are too far exhausted to **admit of** the patient's recovery afterwards. That **delay** tends greatly to diminish the chances of a favourable result from such an operation, is obvious **enough. In a case** which was operated on at the London Hospital on the fifteenth day of obstruction, and ended fatally, I found, on examination of the body, the peritoneal coat of the transverse colon ruptured to the extent of about six inches.

The operation of colotomy may be required not only on account of the difficulty experienced in obtaining **relief from the** bowels, **but** also, as I have mentioned in a previous chapter, in consequence of the extreme misery produced by the disease. Last year I performed colotomy in Dublin on an eminent member of the profession and an esteemed friend, at his own request, to remove the distress consequent on the passage of fæces from the colon into the bladder. He died exhausted by his disease three weeks afterwards, but **he** remained during that period entirely free from his previous sufferings. **In two cases of obstruction from** carcinomatous **stricture,** in which I opened the colon in the loin, the patients' sufferings were aggravated by the formation of an opening between the bowel above the stricture, and the bladder. In a peculiarly distressing case recorded by Mr. Pennell,[7] a communication having formed between the rectum and bladder, and urethra, in which passage, also, there was an impassable stricture, so much irritation and mischief resulted, that the patient gladly submitted to colotomy for relief. In treating of cancer I have remarked that under certain circumstances, although no obstruction exists, it may be our duty to propose colotomy to save the patient much of the misery consequent on the disease during the remainder of life. The danger attending the operation of colotomy is much less than is commonly supposed, and in the unsuccessful cases which have fallen under my notice, the fatal result was chiefly attributable to the effects of the disease of the bowel on the constitution of the patient.

Operations for the formation of an artificial anus have been performed in so many instances with a satisfactory result, that the operation is now regarded as established, and creditable to surgical art. An opening into the descending colon in the lumbar region, external to the peritoneum, is generally admitted to be the best operation. The inguinal, in which the peritoneum is opened, and an anus formed in the sigmoid flexure, is now very rarely performed.

The descending colon may be opened in the left loin in the following manner. The patient is to be placed upon his right side, but inclined towards the face with his back to the window, on a couch or nar-

---

[7] Medico-Chirurgical Trans. vol. xxxiii.

row table of convenient height, and a pillow is to be placed beneath the lower part of the abdomen, in order to render the left flank prominent. The surgeon, standing in front of the patient, so as not to intercept the light, may mark with ink the spot where the intestine is to be sought for and opened. This is about two fingers' breadth between the anterior and posterior spinous processes. Anæsthesia having been produced,[a] an incision is to be made across the loin, commencing at the outer margin of the erector spinæ and carried outwards for about four or five inches. The layers of muscles are to be cut through down to the transversalis fascia, which is to be divided upon a director. In the loose fat beneath this fascia will be found the posterior wall of the colon. When no fat is present, the fascia glides over the bowel and closely resembles the serous membrane, and this is very apt to puzzle the operator. The colon when exposed may often be recognized by a longitudinal band. It is to be seized with the forceps and drawn towards the outer wound, and an incision is then to be made into it in the longitudinal direction. This opening should not be less than an inch in length. Its sides are to be secured to the lips of the wound in the skin by four sutures of stout silk, two on each side. This is an important step in the operation, as it prevents fæcal effusion into the loose areolar tissue, renders the intestinal opening superficial instead of at the bottom of a deep wound, and obviates any after-difficulty in

---

[a] I have entirely discarded chloroform in this, as in most other operations. I am satisfied that I lost two of my patients after colotomy, from exhaustion consequent upon persistent vomiting produced by chloroform.

keeping the new anal aperture patent. A branch of one of the lumbar arteries may be wounded, and may require to be tied. The abdominal branch of the last dorsal nerve, and the trunk of the musculo-cutaneous behind, are also generally divided.

Mr. Bryant performs the operation by an oblique incision, which he prefers, as giving more room for manipulation when the colon is empty, and as taking the line of the nerves and vessels, and so lessening the risk of their division. He makes an incision, four or five inches long, beginning an inch and a half to the left of the spine below the last rib, and passing downwards and forwards in front of the anterior spine of the crest of the ilium, with the line of incision passing obliquely across the external border of the quadratus lumborum muscle about its centre. By this incision the integuments and fascia are to be divided, and the outer border of the quadratus lumborum exposed. The abdominal muscles are then to be divided upon a director.'

Colotomy in the loin is an easy operation in spare emaciated persons with the colon fully distended. But in other conditions there may be great difficulty in opening the bowel external to the peritoneum. In a very fat person the lumbar colon is at a great distance from the surface, and can be reached only by a large and deep incision. I have experienced considerable difficulty in finding the bowel under these circumstances, and I am informed that cases have occurred in which the operator has failed to meet with it, and after great efforts has been obliged to abandon the operation. When the bowel is thus

' Bryant, Practice of Surgery, 2nd edit., vol. i. p. 681.

deeply seated, it may be rendered prominent by an assistant firmly compressing the abdomen. Coughing will occasionally bring it towards the surface. It is a curious circumstance, that in obstructions even of long continuance, the descending colon is sometimes contracted, the fæces having accumulated in other parts of the intestinal canal. This contracted state of the bowel causes great difficulty in the operation. In a case of carcinomatous stricture in the rectum, in which I performed colotomy, on a woman aged forty, after a month's obstruction, not only was the colon contracted, but it was actually compressed against the spine and put out of the way by the distended small intestines, so that it was impossible to reach the bowel without opening the peritoneum. No inflammation or unfavourable symptom resulted. In painful disease of the rectum without obstruction, the surgeon may facilitate the operation by passing a long flexible tube into the colon, and throwing up a large quanity of fluid, such as linseed-tea or gruel, so as to make sure of a distended bowel at the part to be opened. This I have done in several instances with great advantage.

In the performance of lumbar colotomy, the surgeon should be careful to make a good-sized opening in the bowel, to draw it to the surface, and attach the edges of the opening with sutures to the outer wound. If this be neglected, the colon is very apt to recede, and the operation to be followed by contraction of the aperture, so that great difficulty may be experienced in maintaining a sufficient vent, and the fæces passing into the bowel beyond, the object of the operation may be defeated, as in a case related by Dr. Laffan.[1] After the operation, the only application I make to the wound is a piece of spongio-pyline, moistened with a lotion of tincture of iodine (ʒiv—ʒviij) or of a solution of Condy's fluid. This soaks up the feculent discharges, and, being renewed, keeps the part tolerably clean and sweet. The lotion may be injected also into all recesses and sinuses.

An artificial anus in the loin, well established, is attended with much less inconvenience than might be supposed. In a healthy state of the alimentary canal above, fæces pass only at regular intervals, and the escape of flatus and feculent matter at other times may be to a great extent prevented by a well-adjusted pad and bandage, which will also restrain a tendency to prolapsus, which sometimes exists when the opening is quite free. I was greatly struck with the moderate inconvenience consequent on an artificial anus in the loin in the case, already alluded to, of a gentleman successfully operated on by Mr. Pennell, in Rio Janeiro, in 1849. This gentleman was afterwards much engaged in business, and in 1854 came over to this country, when I had an opportunity of seeing and examining him. He died in 1862 of albuminuria, having survived the operation thirteen years. I shall have occasion to allude further to the conditions of an artificial anus in the loin in treating of the operations required for congenital imperfections of the rectum.

No important parts are divided in colotomy, and in persons not too enfeebled by disease a favourable result may be anticipated.

[1] Dublin Journal of Medical Science, Oct., 1872.

I have performed the operation eighteen times, and taken part in three other operations.[2] Fourteen of these patients recovered. In two of the seven fatal cases, the unfavourable result was chiefly due to chloroform. In a case of thirty days' obstruction peritonitis was already set up, and I performed the operation with great reluctance, in deference to the opinion in its favour expressed at a consultation on the case. Another died on the third day also from peritonitis springing from the cancerous disease in the rectum. I lost one patient from pyæmia, and two, in whom obstruction existed, died of exhaustion, one on the sixth day, and the other on the twelfth.

CHAPTER XVII.

ATONY OF THE RECTUM.

In paraplegia the forces which expel the fæces and the retentive functions of the sphincter are both destroyed; consequently the motions, if sufficiently liquid, on reaching the lower bowel escape involuntarily. I have not met with any well-marked case of paralysis of the rectum independently of palsy of the lower half of the body; but several instances of loss of tonicity, or defective muscular power in the lower bowel, rendering it incapable of properly extruding its contents, have come under my notice in practice. The following is a type of such cases:—

[2] Most of these cases have been recorded in detail in the London Hospital Reports, Medical Journals, and third edition of this work.

Case 32.—Mr. K——, aged seventy-three, a retired officer in the marines, who had served a good deal in warm climates, consulted me in consequence of a want of power to empty his rectum. He had suffered from this annoyance more or less for twenty years, and he ascribed its origin to an attack of dysentery. A desire for evacuation came on regularly, but the action occupied a long time, and was a daily misery to him. He was obliged to dislodge all lumps by the introduction of his finger. On examination, I found a relaxed baggy condition of the lower bowel, and some enlargement of the prostate gland. The sphincter yielded readily. I prescribed pills containing extract of nux vomica and watery extract of aloes, but they were not of much service. He derived most relief from mild saline aperients, which, by softening the evacuations, facilitated their passage.

In this instance the sphincter contracted normally, so that the case is quite different from the cases alluded to at page 16, in which the expulsive powers of the bowel were insufficient to overcome the resistance offered by an irritable and hypertrophied sphincter. An atonic condition of the bowel may be produced by the too free and frequent use of enemata, the quantity thrown up being so large as to dilate the bowel and impair the power of its muscular coat. And this may account for certain appearances which have sometimes attracted my notice in travelling on the Continent, viz. ugly marks on the side walls of the closets, even in good hotels, caused by the wiping of the finger which had been used in assisting defecation, for it is well known that lavements are in more common use on the Continent than in this country.

An atonic condition of the rectum is apt to give rise to fecal accumulations. Cases of the kind are not uncommon; yet the nature of the affection is liable to be overlooked by practitioners not alive to its occurrence. It appears that the rectum becomes gradually dilated and blocked up by a collection of hard dry fæces, which the patient has not the power to expel; being **unable, from loss** of tone in the distended bowel, to overcome the resistance **of** the sphincter to the passage of so great a body. Some **indurated lumps from the** sacs of the colon, **on reaching the rectum, perhaps coalesce so as to** form a **large mass;** or a quantity accumulated in the sigmoid flexure, on descending into the lower bowel, becomes impacted there. **In several instances a** plum-stone has been found in the centre **of the mass,** forming a sort of nucleus. Such a collection gives **rise to considerable distress** and alarm, producing **constipation, a sensation of weight** and fulness in **the rectum, tenesmus,** and forcing **pains** which **women describe as** being **equal** in severity to those **of labour. In cases** of some duration, where the **hardened fæces** do not quite obstruct the passage, they excite irritation and a mucous discharge, which, mixing with recent feculent matter passing over the lump, causes the case to be mistaken for a diarrhœa. Injections have no effect in softening the indurated **mass: they act** only on the surface, and return **immediately, there being no room for their lodgment in the bowel. The** surgeon, **on introducing** his finger **at the anus, finds the bowel distended and** blocked up with **a** large lump, which feels almost as hard as a stone. In such cases, the only mode of giving relief is by mechanical interference. The mass requires **to** be broken up and scooped out. For this **purpose a** lithotomy-scoop is a proper instrument; **but, as this is** not always at hand, I have used **sometimes a silver dessert-spoon,** which I have had to pass nearly its whole length, **in order to** dislodge the hardened mass. The surgeon should be content with breaking up and extracting the larger portions, a few injections of soap and water afterwards being sufficient for the removal of the remainder.

Sir James Simpson **has** described this affection under the head of "ball-valve obstruction of the rectum by scybalous masses," and mentions a case in which a large, hard, oblong scybalous mass filled up the canal of the rectum. The mass was broken up and removed piece-meal.[3] The following case occurred in my own practice. **I was asked by her** medical attendant **to** see a lady about fifty years **of** age, who **for eighteen** months had been unable to **relieve her** bowels without aperients and without passing her finger into the rectum. **On** examination I detected **a hard** elongated mass of fæces, which was evidently forced down in the attempted effort of defecation, **and** obstructed **the anus until** the mass was pushed back by the lady's finger. **I** broke up this mass with the handle of **a tablespoon, and the** bowels were afterwards emptied by repeated injections, which entirely removed the difficulty.

I have had to afford assistance in several other cases of this painful and disagreeable affection. No less than three came under my notice during a period of six months. They were all persons enfeebled by age or disease. **One** was the **case** of a lady, **aged** sixty-eight, whose constitutional powers **were much** weakened by long-existing carcinoma of the breast. Her sufferings were so severe that I made the examination in expectation of finding carcinoma of the

---

[3] Edinb. Monthly Journ. of Med. Science, 1849, p. 705.

rectum, or of the uterus, preventing the passage of the fæces. The second was the case of a man, aged forty-seven, whose leg I had amputated in the London Hospital a few weeks before, for disease of the tarsus. He had suffered from secondary hæmorrhage, had a bed-sore, and was much reduced at the time. The third was a bedridden old lady, aged eighty-four, who had taken largely of laudanum for a nervous affection of the throat. They were all readily relieved, by mechanical aid, from a state of considerable suffering and distress.

CHAPTER XVIII.

ANAL TUMOURS AND EXCRESCENCES.

BESIDES the flaps and folds of integument consequent on external piles, other growths are developed in the immediate vicinity of the anus. Thus, tumours of a fibrous texture sometimes form in the subcutaneous areolar tissue, and, as they increase, become pedunculated. They seldom exceed the size of a chestnut. They have a firm feel, and their surface is generally irregularly lobulated. Mr. Hovell, of Clapton, sent me an unusually large tumour of this kind, which he had excised from a gardener, forty-one years of age: it weighed upwards of half a pound, and was composed of fibrous tissue arranged in several lobes: it had been pendulous, and attached to the margin of the anus by a narrow neck. There was an ulcer on its surface, produced, no doubt, by pressure in sitting, and friction against the dress. This tumour had been seven years in forming. Few persons would allow a tumour to increase to such a size, in so inconvenient a situation, without seeking for relief from an operation. These fibrous growths may be easily and safely removed by excision.

Warts are not unfrequently developed around the anus, and they sometimes grow so abundantly as to constitute a considerable cauliflower-looking excrescence. They then form projecting processes, of various sizes, densely grouped together, many being of large size, with their summits lobulated, expanded, and elevated on narrow peduncles more or less flattened. I have removed a mass forming a tumour as large as the closed fist, separating the nates, and almost blocking up the passage for the fæces. When abundant, they are attended with an offensive, thin discharge. They originate in the irritation consequent on want of cleanliness, and occur generally in young grown-up people of both sexes. I once saw a large crop of these growths in a child only four years of age. In some persons there is a strong disposition to the formation of warts; so that, without great attention, it is difficult to prevent their growth. If few in number, and small in size, they may be destroyed with strong nitric acid. They generally require, however, to be removed by excision, which is the quickest and most effectual mode of treatment. This may be effected with a pair of scissors curved on the flat. The operation is painful, and should therefore be performed whilst the patient is under the influence of chloroform. Wet lint may be applied to the part; and the patient should be directed afterwards to check any tendency to a reproduction of the growths, by great cleanliness, and the use of a lotion of the oxide of zinc.

Warty growths around the anus may ulcerate and degenerate, and become the seat of epithelioma.

Case 33.—A lady, aged sixty-one, who had long been subject to excrescences around the anus, was sent to me from the country, on account of the severe pain which she had suffered in these growths for a period of two years. On examination I found some of them ulcerated, and a granular indurated condition of the parts around, with a thin ichorous discharge. The growths were all freely excised, and on examination presented the characters of epithelioma. The part healed with some amount of contraction. Six months afterwards I had to remove a further small painful growth, since which the patient has remained well.

CHAPTER XIX.

ORGANIC CONTRACTIONS OF THE ANUS.

In treating of the irritable sphincter muscle, I noticed an impediment in defecation which sometimes arises from spasmodic contraction at the anus. A more serious difficulty may occur from organic changes diminishing this aperture. Thus in describing the operation for internal hæmorrhoids, I remarked on the contractions which sometimes take place after too free excision of the skin at the margin of the anus, and also after the unskilful use of the écraseur. In some of the cases which have been badly operated on, it has been necessary afterwards to dilate the contracted opening with a bistoury, and to persevere in the use of bougies for some time, in order to maintain an adequate vent for the fæces. After operations for epithelial cancer at the anus, more or less contraction of this part takes place, and may even require mechanical treatment, though it is remarkable how small an opening is sufficient for the purposes of life. This is well shown in the following case :—

Case 34.—A gentleman, of middle age, called on me one day in the autumn, complaining of inability to pass his stool, and of great pain from some obstruction at the anus. On examination I found a hard, rigidly-contracted anus, scarcely capable of admitting the point of the little finger, and a solid body impacted in the opening. Grasping this with a pair of forceps, and using some force, I extracted a plum-stone. On inquiry, I learnt that the patient had been operated on about two years before by Mr. Guthrie, for some kind of growth at the anus, since which the orifice had remained contracted. His evacuations had been small in calibre, but he had experienced no difficulty in passing them, and previous to the obstruction described had not been troubled in any way.

Contractions are also liable to arise after the healing of large ulcers, especially after sloughing phagædenæ of the buttock reaching to the anus. The following is a case in which a serious narrowing of the aperture resulted from a strumous ulcer.

Case 35.—W. S——, a strumous-looking boy, aged four, with light hair and blue eyes, was admitted into the London Hospital, May, 1859, on account of an ulcer with contraction of the anus. The sore was hard, and unhealthy-looking, and nearly surrounded the anus, which was so small that only a No. 1 bougie could pass through it. The bowels seldom acted without castor oil, and were never thoroughly emptied, a quantity of hardened fæces being retained. The bowel having been well cleared out by injections of olive oil, chloroform was given, and I then made an incision from the margin of the anus towards the coccyx, so as to enable a No. 3 bougie to pass. Wet lint was kept applied between the lips of the wound. Cod-liver oil was given three times a day, and castor oil occasionally. The child was kept in bed, and a bougie was introduced daily. The parts about the anus became softer, and the sore began to heal, but after a month I found it necessary to divide a small fistula in front of the anus, communicating with the rectum. The wound quite healed in six weeks after the first operation, a No. 6 rectum bougie could be passed without difficulty, and the retentive powers of the sphincter were preserved. At this time the appearance of strumous ophthalmia, and a failure in the general health, led me to send the child into the country.

CHAPTER XX.

PRURIGO ANI.

ITCHING at the anus is a common symptom in several disorders of the lower bowel; but it may also occur as a distinct affection, or independently of any other disease of the part, being due to a peculiar hyperæsthesia of the skin.

Prurigo ani is caused by worms in the lower part of the rectum, and also frequently results from the determination of blood which attends the formation of hæmorrhoids. The congestion of the hæmorrhoidal veins occurring in chronic enlargement of the prostate gland is also sometimes attended with the same symptom. When the complaint is dependent on piles, and indeed generally, patients suffer mostly after taking wine or stimulating drinks, and during warm weather, and when heated in bed. The itching is most teasing and annoying, but especially at night, when it keeps the patient awake for hours. Rubbing the part to arrest the irritation only aggravates the mischief afterwards; yet few persons have sufficient self-control to prevent their seeking temporary relief by friction; and some, though capable of restraining themselves whilst awake, fret the part unconsciously during sleep. The friction thus resorted to excoriates the skin at the margin of the anus, so that, in chronic cases, the skin becomes dry, harsh, and leathery, loses its pigment in places, presenting dull white patches, cracks from slight causes, and ulcers and fissures are produced, which are but little disposed to heal.

In women, itching at the anus is sometimes consequent on affections of the womb.

Case 36.—I saw, with Mr. Kennedy, of Stratford, a lady who had retroversion of the uterus, which probably produced congestion of the hæmorrhoidal veins. Her most distressing symptom was excessive prurigo, which affected not only the verge of the anus, but even the mucous membrane inside the sphincter. The margin of the anus was excoriated and fissured by friction, and the mucous membrane of the rectum was rough and granular from the same cause, for the sufferer was in the habit at night of inserting her finger within the bowel to endeavour to allay the tormenting irritation.

In most instances this complaint, after proving troublesome for an hour or two at night, and in the day after excitement, ceases, and the patient has long intervals of rest and ease. But in the worst forms of the malady, the torment is most distressing.

It lasts throughout the night, so that the patient gets little but broken sleep, and after a time the general health seriously suffers, and life is rendered truly miserable. Such was the condition of the lady whose case I have just alluded to.

In some of the cases which have fallen under my notice I could discover no local cause whatever to account for the prurigo. It seemed to be purely an affection of the nerves of the part. The patients were generally healthy persons. One gentleman, who had been subject to it for years, found that it was connected with his state of mind. When much engaged, and prosperous in business, he suffered little from it. He was sometimes free from it for a whole month, and then became troubled for many nights in succession. In cases of this kind, the complaint is usually very obstinate, and sometimes severe; but after proving more or less troublesome for years, it has been observed to subside as age advances.

In prurigo, by whatever cause produced, the habits of living should be regulated. The patient should sleep on a mattress, and be as lightly covered as is consistent with comfort. Cold bathing or sponging should be daily resorted to, and sufficient exercise taken in the open air. All hot condiments and stimulating drinks must be strictly avoided. The actions of the bowels are to be regulated, if necessary, by medicine; and after each evacuation the parts should be well cleaned with soap and water. Every effort must be made to avoid friction; and the patient should be assured, that if he yields to his inclinations his complaint will be rendered worse, and more difficult of cure.

In prurigo from obvious local congestion, a leech or two at the anus will give marked relief. In all cases, the disease which gives rise to this troublesome symptom must be the chief object of attention; but there are certain remedies which are specially adapted to relieve the irritation. The itching attendant on piles may generally be arrested by smearing the anus with some mercurial ointment, as the dilute citrine, or one containing the grey oxide of mercury (ʒj.—ʒj.), or by lodging in the aperture a small piece of cotton wool soaked in a lotion of the oxide of zinc (ʒj.—ʒviij. of water and glycerine). This lotion is sometimes rendered more efficacious by the addition of the dilute hydrocyanic acid. Lotions of carbonate of bismuth and glycerine, of borax and morphia, or of carbolic acid are often efficacious in this complaint. The application to the anus of a strong solution of nitrate of silver (gr. xx.— ʒj.) with a camel's hair brush once daily often gives relief, especially in cases where the skin is made rough and sore by rubbing. In some cases great relief may be derived from chloroform ointment. It produces a smarting sensation when first applied, but this is soon followed by ease. In persons of weak constitution, benefit has resulted from full doses of quinine, and in certain cases liquor arsenicalis with steel, has helped to relieve the irritation. The complaint is often very obstinate, and much perseverance is required on the part of the practitioner, and also of the patient, to effect a cure.

CHAPTER XXI.

CONGENITAL IMPERFECTIONS OF THE ANUS AND RECTUM (ATRESIA).

Of the many congenital imperfections and deficiencies to which the body is liable, there are few more fatal to the life of the infant, or more distressing to the parents, than those of the lower bowel. They are very varied in form, and the opportunities of observing them occurring but seldom, it is scarcely surprising that a satisfactory description of these defects is rarely to be met with in surgical works, and that no clear rules are laid down to guide the practitioner in the treatment of them. It will be found on inquiry, that in a large number of instances the operations performed to remedy these imperfections have failed in procuring a vent for the fæces; that in many others, where a passage has been made, the relief, though immediate, has not been sufficient or permanent, but has been followed by contraction, and has only prolonged a miserable existence; whilst in a very few only has success been complete, and life been preserved without risk and serious inconvenience. With the view of ascertaining the more common forms of these affections, and of helping to determine the best mode of dealing with them, either for the preservation of life or its future comfort, I collected, tabulated, and analyzed one hundred cases, in which operations had been performed by myself and other surgeons, and I communicated the results to the Royal Medical and Chirurgical Society in 1860.' Additional cases have since fallen under my notice, and I hope now to give a succinct but tolerably complete account of these imperfections, and to indicate the treatment applicable to them.

The congenital malformations of the rectum may be conveniently classed as follows:—

1. Imperforate anus without deficiency of the rectum.

2. Imperforate anus, the rectum being partially or wholly deficient.

3. Anus opening into a cul-de-sac, the rectum being partially or wholly deficient.

4. Imperforate anus in the male, the rectum being partially or wholly deficient, the bowel communicating with the urethra or neck of the bladder.

5. Imperforate anus in the female, the rectum being partially deficient, and communicating with the vagina.

6. Imperforate anus, the rectum being partially deficient, and opening externally in an abnormal situation by a narrow outlet.

7. Narrowness of the anus.

A few other congenital deviations have been observed, but they are of very rare occurrence, the seven forms enumerated above being those more commonly met with. The following case occurred in my own practice:—

Case 37.—A male child, five weeks old, was brought to me on account of a congenital fecal fistula communicating with the rectum. There was a depression at the back of the sacrum, about an inch and a half from the anus, with a small

' Transactions, vol. xliii.

orifice, through which a little feculent matter escaped, when discharges occurred from the natural orifice, which was of proper size. The fistula was too small to admit the passage of a common-sized probe. It closed spontaneously before the child reached two years of age.

According to my table these malformations occur more commonly in male children than in female. Thus sixty-eight were males, and thirty-two females. This does not agree, however, with the observations of M. Bouisson, who also collected 100 cases, of which fifty-three were females, and forty-seven males.[5] I am unable to account for this discrepancy. In my table, which excludes the first and seventh forms of malformation, there are twenty-six instances of the second form, thirty-one of the third, twenty-six of the fourth, eleven of the fifth, and six of the sixth. The fourth is the most common form in the male sex, and twenty-six is probably less than the actual number, for the communication with the urethra is sometimes so minute as to prevent the escape of meconium during life, and readily to escape detection after death.

It will be observed that the classification of these imperfections is founded on states which can generally be recognized during life. Unfortunately the condition of the terminal portion of the intestinal canal, and its relations to the parts around, cannot be predicated with any certainty. In cases of imperforate anus, or of anus opening into a cul-de-sac,

---

[5] Thèse sur les Vices de Conformation de l'Anus et du Rectum, p. 73. The late M. Giraldès, Surgeon of the Hôpital des Enfants Malades, in Paris, found these imperfections more frequent in boys than in girls.

the intestinal canal may **terminate in** a blind pouch at the brim of the pelvis, the **rectum being** wholly wanting; or an imperfect **rectum may form a shut** sac, descending to the floor of the pelvis, or **as low** as the neck of the bladder in the male, or the com- **mencement of the vagina in the** female. It is known that the anal portion of the bowel is developed distinctly from the **upper** portion, and that the two afterwards approximate and unite, the diaphragm or septum disappearing by interstitial absorption. A failure in this process is the cause of the **second** form of congenital imperfection. The **cases of im**perforate anus in which the rectum communicates with the urethra or vagina depend on the original existence of a cloaca, the malformation being due to an incomplete **separation during fetal life.** These conditions are the result of an arrest of development at different stages. The blind pouch, in which the intestinal canal terminates, is sometimes connected to the anal integument, or **to the anal cul-de-sac, by a cord** prolonged from the **bowel above.** These cases are not, like the preceding, the result of a non-formation of the rectum, but are produced by an obliteration of the bowel, which was originally well formed; the obliteration being a pathological change, due probably to ulceration and adhesion which had taken place during intra-uterine life. M. Goyrand, who has published **some able** papers on these malformations in the "Gazette **Médicale de Paris**" (1856), quotes **a case** communicated by **M. Forget to** the Société Médicale d'Emulation, **of** a female **infant** with **an** imperforate **anus, who** died after an operation in the perineum, without the bowel being reached. The intestine terminated at the base of the sacrum in a pouch, from which proceeded a good-sized cord to be implanted in the skin at the site of the anus. This cord was formed of fibres continuous with the longitudinal fibres of the muscular tunic of the intestine. Friedberg also mentions a case in which the obliteration was less complete than in the preceding. It was that of a female with an anus well formed, but with the rectum closed at a short distance above. At the autopsy the walls of the intestine were found adherent to each other in different places.[6] In a case, recorded by Mr. O'Connor, of imperfect rectum, the anus opening into a cul-de-sac, the commencement of the descending colon suddenly contracted into a fibrous cord which terminated near the promontory of the sacrum.[7]

The relations of the peritoneum to the bowel in the different forms of atresia have an important bearing on the operations performed in the perineal region; for there can be no doubt that, in many instances, a fatal result was due to an opening made in the serous sac. When the rectum, being only partially defective, descends into the pelvis, as commonly happens in the fourth, fifth, and sixth forms, the peritoneum is reflected from the bowel at a distance from the part liable to be opened in the ano-perineal operation. But when the rectum is wholly wanting, the intestine not passing lower than the brim of the pelvis, the peritoneum completely invests the terminal pouch, and must necessarily be wounded before the bowel can be reached

[6] Archives Générales de Médecine, Ve Série, t. ix. p. 569.
[7] British Medical Journal, Dec. 1860.

from the perineum. In cases of intra-uterine obliteration of the rectum, the peritoneum is also much exposed to injury. In Forget's case, quoted above, the serous membrane extended upon the **fibrous cord for three lines, and was** strongly adherent to it.

In cases of complete deficiency of the rectum, it has **been remarked** by Rokitansky, Goyrand, and others, that the pelvis is not well developed, the tuberosities of the ischium being near together, **and** the antero-posterior diameter abnormally small. A depression in the anal region, and **the position of** the genitals far back, **would also** lead us to infer an absence of the rectum. In some instances the sacrum has been found defective and the coccyx wanting.

1. *Imperforate anus without deficiency of the rectum*, is the simplest form of congenital malformation producing complete obstruction. The occlusion is **caused by a prolongation of the skin over** the terminal extremity of the bowel, **and the barrier is** usually **so thin and superficial, that the meconium** can be distinguished by the black or dark blue colour of the skin. This impediment can be readily and effectually removed. A central crucial incision should be made with a sharp-pointed bistoury, and the prominent angles of **the** integument excised. A bougie must be **passed daily until the part is** quite healed and the aperture is fully established, **which seldom occupies more than a** week.

2. *Imperforate anus, the rectum being partially or wholly deficient.*—When partially defective, the rectum is rarely **found** at a less distance than an inch **from** the perineum. In entire absence of the lower

bowel, the colon generally terminates at the brim of the pelvis, but even this portion of the gut may be more or less defective. In this form an early attempt should be made to open the bowel in the ano-perineal region. The surgeon should cut down at the site of the anus, exactly in the median line, extending his incision towards the coccyx, dividing the musculo-aponeurotic floor of the pelvis, and penetrating to the depth of an inch and a quarter; and then, if no bowel be reached, the operation should be suspended or abandoned. If no long period have elapsed since birth, and if the infant be not exhausted by sickness or want of nutriment, the surgeon can wait eight or twelve hours, and again examine the wound; for it may happen that the rectum, forced down by the infant's struggles, and finding no resistance from the floor of the pelvis, will project between the borders of the incision.[s] If then any protrusion, or even bulging, be observed, the swelling may be explored with a grooved needle or the point of a bistoury. After the surgeon has reached the bowel, and opened it freely, its coats should be grasped with the forceps, and gently drawn down to the external wound, to the margins of which they should be attached by two or more

---

[s] Petit records a case operated on by two surgeons. The first made a conical incision in the perineum, but did not succeed in reaching the rectum. Another surgeon, who saw the infant three hours afterwards, was surprised to find a dark swelling projecting at the wound, and concealing it. This tumour was incised, and meconium escaped. The infant died in seven or eight days, and at the autopsy Petit found that the black tumour was formed by the posterior part of the rectum protruded by the efforts of the infant at the part where the least resistance offered. Mémoires de l'Académie Royale de Chirurgie, t. i. partie ii. p. 210, 1743.

silk sutures. Care must be taken not to use force, so as to tear the tender walls of the bowel with the forceps, or to cause the sutures to cut through them. With the object of avoiding this,—of more readily reaching the bowel, and of gaining more space for drawing it down, Verneuil has recently revived an important addition to the operation originally suggested and practised in one case by Amussat,[9] viz:—resection of the coccyx, which Verneuil performed in six cases.[1] The point only of the coccyx is to be taken away by prolonging the median incision and detaching the parts laterally with the scissors. The attachment of the external sphincter must be divided in this part of the operation, but is not likely to cause any permanent weakness. The wound must afterwards be poulticed and kept clean, and the sutures may be left in for five days or a week. After a week or ten days a bougie should be passed occasionally, for several weeks at least, even in favourable cases, to obviate any tendency to contraction. When the perineum is rendered convex during the infant's struggles, the operation is pretty sure to be attended with success; but when this part is depressed, and the pelvis obviously small and narrow, the prospect is so unfavourable, that it is question-

[9] Sur la Possibilité d'établir un Anus artificiel, &c. Troisième Mémoire.

[1] Resection de Coccyx pour faciliter la Formation d'un Anus périnéal dans les Imperfections du Rectum. Paris, 1873. Verneuil excised the coccyx in all six cases with satisfactory results as respects the operation. One patient was going on well at the age of nine years, and a second at the age of four months. The others died from various causes within two months. Esmarch disapproves the operation, as he says the same effect may be obtained by forcibly bending back the coccyx.

able whether any attempt should be made to obtain a passage in the anal region. In this case, or in the event of a failure to reach the bowel, the operator should state the circumstances to the **parents, and, with their consent,** open the colon in the **left groin.**

3. *Anus opening into a cul-de-sac, the rectum being partially or wholly deficient.*—The anal pouch varies in depth from half **an inch to an inch and a half. Its fundus may be separated** from the blind **extremity of** the rectum by a septum composed only **of the coats** of the two portions of the bowel and a little areolar tissue, or the upper part of the rectum may be want**ing altogether or** nearly so, a space more or less con**siderable intervening.** The upper and lower portions **of the** rectum usually **meet end** to end, but this is **not** constant; **for in a** case observed by myself, in one described **by** Amussat,[2] and in another recorded by Godard,[3] the lower portion ascended in front of the upper, so that for a short distance the two passages ran parallel. In Amussat's case this arrangement was favourable to the result of the operation by incision into the bowel above, and division of the **septum between it and the anal cul-de-sac.** In Godard's **case merely a puncture was resorted to; consequently the** collection of **fecal matter in the upper or posterior pouch** pushed the partition forwards, so that it formed **a valve, which** prevented the escape of the meconium without the introduction of a tube, a difficulty which soon led to a fatal result.

In a case of this malformation, an attempt should

[2] Lib. cit.   [3] Gaz. Méd. de Paris, 1856.

be made at once to reach the terminal portion of the intestinal canal. The cul-de-sac might, if the symptoms are not too urgent, be first dilated by a sponge tent for a few hours. If the surgeon should feel the upper part of the distended rectum projecting into the lower, he might explore the swelling with a grooved needle, and if gas or meconium escaped, make a free opening with a bistoury. He is liable, however, to be deceived by a distended bladder filling up the pelvis. Giraldès in a case of the kind, on introducing his finger per anum and detecting a fluctuating swelling, passed an exploring trochar, and gave issue to a large quantity of urine. If nothing be felt to indicate the near proximity of the bowel, the surgeon should enlarge the anus by an incision, carried towards the coccyx, so as to divide the posterior wall of the sac, if necessary excise or bend back the point of the coccyx, and should pursue his search in this direction to the depth of an inch and a half or two inches from the anus. If the bowel be not reached, this operation should be abandoned, and colotomy performed in the groin. There is seldom any use in delaying the latter operation, in expectation of the bowel becoming more loaded, and descending into the perineal wound, as this malformation is rarely detected until obstruction has already existed for some days. After an opening has been made, if it be obvious, from the depth of the incision, that an interval of some extent exists between the two ends of the bowel, the upper portion must be drawn down and secured to the wound. When a mere septum intervenes, this is not necessary, but careful and even long-continued dilatation will be required afterwards as a security against contraction.

The plan of drawing down the bowel and securing it by sutures to the margin of the wound made in the skin was, I believe, practised first by Amusat, in 1835, in a case to which I shall presently refer. This mode of proceeding has also been particularly described and recommended by Dieffenbach.[4] The important advantage obtained by it is the securing of a lining of mucous membrane for the passage traversed by the fæces. By this means we not only guard against the tendency to contraction, with its consequent miseries and dangers, but also avoid the early risks of inflammation of the areolar tissue and peritonitis. This plan of operating was adopted by me, in the following case, with a most satisfactory result :—

Anus opening into a shut sac, the rectum being imperforate and partially defective. Operation and permanent cure.

Case 38.—A male infant, five days old, was brought to me at the London Hospital, April 16th, 1860, in consequence of sickness and inability to pass the fæces. He was weak and emaciated, and had not slept for twenty-four hours. Several doses of castor oil had been given to him. The anus was well formed, but on passing a director and afterwards my finger, I found it to open into a cul-de-sac, three-quarters of an inch in depth. A piece of sponge having been inserted and kept in for half an hour, in order to dilate the sac, I enlarged the anus by an incision towards the coccyx, dividing the posterior wall of the anal pouch to the depth of nearly an inch, when the upper portion of the rectum was opened, and meconium escaped. The bowel was seized with the forceps, drawn down, and secured to the wound in the skin on each side by two sutures. After the operation the infant took the breast well, slept, and passed its motions freely.

[4] Die Operative Chirurgie, 1845-8.

The sutures were removed on the 21st. May 3rd, though defecation was unimpeded, a bougie was passed for the first time, and, as a matter of precaution, directed to be introduced every other day, the size being gradually increased to a No. 4 rectum bougie, which was afterwards passed at intervals increased from a week to a month, until a free passage was quite established. The infant has thrived well, and in January, 1864, was a remarkably fine boy, with the passage quite free.

Mr. Le Gros Clark operated in a case, the history of which illustrates in a striking manner the difficulties of treatment, and the serious inconveniences so liable to arise after the formation, simply by incision or puncture, of a communication between the two portions of the rectum, when an interval of any extent exists between them. The cul-de-sac was only half an inch deep, and in order to reach the upper portion of the bowel he had to carry his knife to the depth, including the sac, of two and a half inches before the gut was fairly opened. To preserve the passage, the thickened and contracted tissues required division at the end of three weeks. For several months the infant suffered from constipation and diarrhœa, with great distension of the abdomen, and about ten months after the second operation he again divided an obstructing band with a hernia-knife. Dilatation was afterwards persevered in. The boy at nine years of age was in tolerable though not robust health, but subject at times to serious troubles in defecation.

M. Amussat has recorded a case which is remarkable from the circumstance of the natural anus opening into the vagina, both portions of the rectum being deficient. In this, so far as I know, unique case, it was an important point, in order to prevent the entry of the fæces into the vagina, to establish a new and distinct passage. Amussat made, therefore, an incision behind the anus, and the rectum being wanting, penetrated deeply (two inches), until he reached the bowel, which he detached, dragged to the perineum, and secured to the skin by sutures. The child subsequently came under the care of Sir Philip Crampton, who had to enlarge the opening at the end of two months. The ultimate result was perfectly satisfactory. The patient grew up and married at the age of twenty-one. Though the operation succeeded, and the patient retained the fæces perfectly well, I agree in the opinion since expressed by Amussat, that in a similar case it would be much more satisfactory to unite the extremity of the intestine to the normal anus.

4. *Imperforate anus in the male, the rectum being partially or wholly deficient, the bowel communicating with the urethra or neck of the bladder.*—In this form the rectum is seldom entirely wanting, but the communication with the urinary passage at the neck of the bladder, or more commonly at the anterior part of the prostatic portion of the urethra, is always small and insufficient, and not usually direct, but takes place by a narrow canal. The meconium escapes consequently with more or less difficulty, and retention occurs at an early period. Liquid meconium may pass readily at first, but as the fæces acquire consistency, obstruction arises and life becomes endangered. There are a few cases on record in which, the communication being more free than usual, life has been preserved for many months, the fæces escaping entirely by the urethra, until, the passage

becoming at length blocked up, death has ensued. This happened in two cases, in which an attempt was made to reach the bowel by operation without success. In one, related by Mr. Williamson,[5] great distress resulted, but the child survived eight months and twenty-two days, death being partly caused by ulceration of the urethra and infiltration of urine. In the other, recorded by Flajani,[6] the infant suffered considerably until the opening into the urethra became blocked up by a cherry-stone, and death took place at the age of eight months. Dr. Steel has recorded a case in which no operation was performed, and the child lived eleven months, when the passage became obstructed by two large apple-seeds, which caused death:[7] and Dr. Lyell, of Dundee, met with a case in which the child lived twelve months before fatal obstruction occurred.[8] The opening into the urethra appears to be usually of a valvular character, so that, although the fæces can pass into the urinary canal, the urine is unable to enter the rectum.

In this form of malformation, although meconium may escape by the urethra, an operation for obtaining a vent at the natural site should not be delayed, or serious inconvenience is likely to result, after the establishment of an artificial anus, from previous over-distension of the bowel. The operation should be performed as in the second form (see page 202), and especial care should be taken that the new channel is made of ample size, in order to afford every facility for the subsequent closure of the com-

[5] Lond. Med. Gazette, vol. xxxvii.
[6] Observationi di Chirurgia, t. iv.
[7] American Journal of Medical Sciences, vol. xv.
[8] Edinb. Monthly Journal, August, 1847.

munication with the urethra. As the rectum is seldom entirely deficient, the bowel is generally reached with greater ease and certainty in this form than in the two preceding, and the results consequently are somewhat less unfavourable. If, however, the bowel be deeply seated, so as not to be found at a depth of an inch and a half from the surface of the perineum, the operator should desist and have recourse to colotomy.

After the establishment of a passage at the anus, the escape of fæces by the urethra is liable to continue, and in several instances serious inconveniences have resulted from non-closure of the abnormal communication, the patients having suffered for years afterwards from painful and difficult micturition. In an interesting case operated on by Dr. Miller, there was great difficulty in maintaining the artificial passage, and repeated operations were performed. The bladder was wounded in the last, ever since which urine was discharged by the anus. A large alvo-urinary calculus gradually formed in the bowel, and was extracted with difficulty by operation at seven years of age.[9] The difficulty of obtaining the closure of the communication with the urethra renders the treatment of this imperfection the least satisfactory of any of the forms of atresia.

5. *Imperforate anus in the female, the rectum being partially deficient, and communicating with the vagina.*—In this form the bowel, after descending well

---

[9] Edinb. Med. Journal, vol. ii. 1856-7. Dr. Marcet (Calculous Disorders, 2nd edit., p. 135) has given an account of a calculus the size of a walnut, composed chiefly of the phosphates found in the rectum of an infant born with an imperforate anus, the bowel communicating with the urinary organs.

into the pelvis, curves forwards and terminates in an aperture, which is sometimes pretty free, but an operation, if not required for the preservation of life, is usually performed to remove a disgusting infirmity. In a woman aged twenty-eight, who was under my care in the London Hospital, and in a girl aged twenty, whom I saw in St. Mark's Hospital, there was no difficulty in defecation, but they had very little power of retention when the motions were lax. In several instances persons born with this imperfection have passed through life, submitting to the annoyances consequent upon it, and have married and borne children. The woman above referred to, who was under my care in the London Hospital for an affection of the bladder, had been married four years, and her husband had never discovered the malformation which she studiously concealed, and reluctantly communicated to me. She had not been pregnant. Morgagni mentions a case of a woman who lived to the age of a hundred years, and passed all her fæces by the vagina. The communication, however, is frequently insufficient, and the bend in the bowel seems unfavourable to the free escape of its contents. In a case which fell under my own observation, there had been occasional constipation; and on examination of the child, who died at the age of four years, a month only after the operation for an artificial anus, a very dilated and hypertrophied rectum fully proved that an impediment had existed during life. In other cases on record, operations have been urgently called for several months after birth, in consequence of the infants suffering from tenesmus and obstinate constipation. An immediate operation is not required; but as the fæces acquire consistency difficulties arise, and in most cases it will be safer not to delay measures for securing a free vent longer than a month.

Three different operations have been practised to remedy this imperfection—1, the enlargement of the vaginal outlet; 2, the establishment of a new passage at the natural site, and closure of the abnormal anus; and 3, detachment of the rectal outlet from the vagina and its establishment in the natural site.

1. Vicq-d-Azyr first suggested the enlargement of the original outlet, by division of the posterior wall of the vagina and the perineum as far as the coccyx, and retaining a canula in the bowel. An operation of the kind was performed by Dr. Barton of Philadelphia. In an infant aged nine months, who experienced great difficulty in defecation, he divided, upon a director introduced at the vaginal orifice, the posterior wall of the vagina and the integuments as far back as the site of the natural anus. Dr. Parish, of the same city, performed a similar operation on an infant fifteen months old.[1] A route was thus established direct from the rectum; but as there could have been no perineum, the anus if not opening into the vagina being in the closest proximity to it, the result cannot be viewed as quite satisfactory. Dieffenbach improved upon this operation. After making a similar but freer division of the vaginal septum and perineum in an infant three months old, he dissected round the rectum so as to isolate it from the surrounding parts, and, drawing

---

[1] These two cases are quoted from the Medical Recorder by Dr. Bodenhamer, in his Treatise on Congenital Malformations of the Rectum and Anus, published in New York in 1860, p. 268 and seq.

it down, attached its edges to the cleft perineum. Three weeks afterwards, when the parts had healed, he formed an artificial perineum. He commenced by separating from the vagina the superior wall of the rectum, which receded about half an inch. The parts below were refreshed and brought together by two hare-lip pins and the twisted suture. The cure is reported to have been complete.[2] I have had no experience of this mode of operating.

2. A new passage at the natural site may be made by passing a curved director or sound through the recto-vaginal opening into the bowel, with its point directed to the site of the anus, and cutting upon it in the median line, care being taken to carry the incision backwards towards the coccyx and away from the perineum, in order to preserve as wide a barrier as possible between the abnormal and artificial apertures. The opening into the bowel should be free, and its coats should be secured by sutures to the borders of the external wound. In a case operated on by Dr. Sharpless, of Philadelphia, and related in an early volume of the "Lancet," the vaginal opening is reported to have closed spontaneously two months after the operation in the perineum. So fortunate a result is very unlikely to occur, and could happen only when the abnormal aperture is unusually small. A plastic operation will generally be required some time after the establishment of the anus in the natural site. In two cases which have fallen under my notice the artificial anus had been made too near the original outlet, the intervening septum being so

---

[2] Quoted by Dr. Bodenhamer from Necker's Annalen, vol. iii. 1824.

slight as not to admit of an operation for closing the original aperture. **In both** of them the fæces escaped freely by the vagina, as well as by the artificial opening. **After a** successful operation in the **perineum,** no attempt should be made to close the **recto-vaginal opening** until the artificial anus is fully established, **and all dilatation has** ceased to be necessary. If the abnormal aperture be small, its contraction and closure may be effected by **touching** the edges **with the actual cautery; but if it be of** large size, a plastic operation will be necessary. The edges may be pared and brought together by sutures, **the** bowels being kept at rest for several days by **opiates until union is secured.**

The following curious **case is** probably unique, and well deserving of record.

Imperforate anus with a double vagina and bifid uterus, the rectum opening into the right division of the vagina.

CASE 39.—In May, 1872, I saw, with Dr. Owen Rees and Mr. Spencer Wells, a young lady, aged twenty-four, who was born with an imperforate anus, the rectum communicating by a **large** opening with a vagina. An anus in the **natural site had been established by operation soon after birth.** The opening was quite **free, but the perineum was** very slight, and the **rectum had an abnormal curve** backwards. The **outlet of the pelvis was** narrow. **It was** determined that an attempt should be made to close the recto-vaginal opening. The barrier formed by the perineum was first divided, and it was then discovered that there was a double **vagina** and a bifid uterus, and **that** the rectum opened only **into the right division of the** former. The lateral surfaces **were pared** and brought together with sutures, completely closing the recto-vaginal communication, but leaving openings into the two divisions **of the vagina.** Union **seemed to have** taken place after four days, but unfortunately when the bowels were relieved at the end of a week, the adhesions were all broken down by hardened fæces. The patient derived some benefit from the operation, as the division of the perineal band enabled the fæces to pass out more freely, whilst the retentive power was unimpaired.

3. We are indebted to Rizzoli, of Bologna, for skilfully devising the third mode of remedying an imperforation with a recto-vaginal opening.[3] His operation consists in laying open the perineum, dissecting round the rectum, and separating the anterior wall, together with the abnormal opening from the vagina, and bringing the opening into the position of a normal anus, preserving the sphincter intact. An incision is made along the raphé, from the fourchette to the coccyx through skin and areolar tissue, and two lateral flaps, including the muscles of the part, are formed. The rectum is carefully dissected away from the vagina and surrounding parts, and the opening moved down towards the coccyx, and fixed there by sutures. The perineal flaps are brought together in front and secured also by sutures. This operation was first performed on a girl aged nine, in 1856. In 1864 Rizzoli had the opportunity of examining the girl just previous to her marriage. She could expel and retain her fæces regularly, even when liquid, and the appearance of the parts was natural, a firm cicatrix existing at the site of the original opening. Rizzoli performed a similar operation, with modifications rendered necessary by different conditions, such as a contracted passage, and a dilated bowel, in three other cases,

[3] Clinique Chirurgicale, traduit de l'Italien, p. 452.

on an infant aged twenty-six months, an infant aged seventeen months, and a girl aged fourteen years, and in all three with a satisfactory result.

6. *Imperforate anus, the rectum being partially deficient, and opening externally in an abnormal situation by a narrow outlet.*—In the male, the opening may be in the perineum just behind the scrotum, in the scrotal raphé, or anterior to the scrotum. In the female, the opening occurs in the perineum close to the vagina, or at the posterior commissure of the vulva. In both sexes and in all these situations the vent is insufficient, and defecation more or less difficult. The existence of an outlet for the fæces prevents immediate danger to life; but sooner or later serious inconvenience arises. As the fæces acquire consistency, they escape with increasing difficulty. The bowel undergoes excessive dilatation, constipation and retention at length ensue, and life becomes endangered, so that an operation is required at an early period.

Two modes have been adopted for remedying the imperfection,—1, the enlargement of the original outlet; and 2, the formation of a new anus at the natural site.

The enlargement of the original opening has been performed in two cases by M. Goyrand,[4] who strongly advocates this operation in preference to the formation of an artificial anus at the natural site. In one, a male infant, the rectum opened in the perineum, behind the scrotum, by a contracted orifice. At six months the operator enlarged the aperture by an incision carried into the perineum, and afterwards

---

[4] Gazette Médicale de Paris, 1856.

secured the borders of the divided intestine to the cut surface of the skin by sutures. At sixteen years of age defecation was free, and power of retention complete, even when the motions were lax. In the second case, a female, the rectum opened by a narrow orifice at the posterior commissure of the vulva. At the age of eleven months M. Goyrand enlarged the aperture by an incision towards the coccyx, and applied sutures, as in the former operation, with a satisfactory result, but the child died, six months afterwards, of a cerebral affection.

Two cases which have fallen under my notice were also relieved by this operation.

Imperforate anus, the rectum opening in the perineum. Operation. Anus established at natural site, and permanent cure.

Case 40.—In December, 1860, I saw, with Mr. Painter, of Beaufort Gardens, a female child, aged three months, with the anus imperforate, and a small abnormal opening in the perineum, close to the posterior commissure of the vulva. The evacuations passed freely after birth, but in a few weeks defecation became difficult, and this increased until the infant was constantly struggling to get relief. During the straining the perineum appeared rounded and prominent. With a probe-pointed bistoury I enlarged the abnormal anus at once, by a free incision in the perineum, exactly in the median line, and extracted a large mass of impacted fæces with the scoop of a director. In this way, and by the child's own efforts, an extraordinary accumulation was removed. The mucous membrane of the bowel was afterwards attached to the skin on each side of the incision by sutures. A free passage was maintained without difficulty, and in the summer of 1862 the child was in good health.

In January, 1861, I saw at the London Hospital,

with Mr. Gowlland, a male infant, aged two months, who had an imperforate anus, and an abnormal opening at the scrotal raphé near the perineum, through which the fæces passed with great difficulty. Mr. Gowlland performed the same operation as in the preceding case, and with an equally satisfactory result. The chief advantage of this simple operation is security against subsequent contraction: for as the upper angle of the wound is untouched, the lower one only has to undergo cicatrization. There is also no necessity for a second operation to close the abnormal opening, which may be necessary when a new and distinct outlet is established.

The formation of a new anus at the natural site is the only operation which is practicable in cases of imperforate anus in the male, in which the abnormal outlet is situated in the scrotal raphé or in front of the scrotum.

Imperforate anus; rectum opening in front of the scrotum. Operation and anus established. Death from erysipelas.

Case 41.—In October, 1861, I assisted my colleague, Mr. John Adams, in an operation on an emaciated-looking male child, three days old, who was brought to the London Hospital on account of an imperforate anus. There was a small aperture beneath the penis just anterior to the scrotum. Through this opening a little dark meconium escaped. Mr. Adams introduced a fine director, which passed along a narrow sinus into the bulging perineum. He cut upon this in the median line, and made a free opening into the bowel, when a quantity of black fecal matter escaped. The sides of the bowel were afterwards drawn down and attached to the external wound by four sutures. The child was taken home to the parents, who were poor people, and was unfortunately attacked with erysipelas, and died a week after the operation.

In a case of similar imperfection which occurred many years ago to Mr. South, at St. Thomas's Hospital, he made an incision an inch deep in the perineum, and simply opened the bowel. In a fortnight the wound closed, and a fresh incision became necessary. Great difficulty was experienced in maintaining the opening, and the operation was repeated four times. The patient was last seen at the age of eighteen. The perineal anus appears never to have been free, and fæces continued to pass occasionally by the abnormal channel in front of the scrotum.[5]

In a male child in which a sinus opened in the perineum behind the scrotum, Friedberg cut down upon the bowel, detached and drew it down to the external wound, and confined it to the skin by sutures. The child died three months afterwards of pneumonia, but the result of the operation at that period was quite satisfactory.[6]

The following are the particulars of a case of a female infant[7] which I treated by the second mode of operation :—

Imperforate anus; rectum opening in the perineum. Operation, and anus established at the natural site. Attempted closure of the abnormal outlet by a second operation.

Case 42.—In November, 1857, I saw, with Mr. Gardner, of Gloucester Terrace, a female infant, a few weeks old. He had noticed at its birth that the anus was remarkably small, and in advance of its usual situation, and he had found it

---

[5] St. Thomas's Hospital Reports, vol. i., and Chelius's Surgery Trans. by South, vol. ii. p. 329.

[6] Archives Générales de Médecine, Ve Série, t. ix.

[7] A similar case, in which an artificial anus was established in the anal region, is related by M. Berard, Gazette des Hôpitaux, 1844, p. 286.

necessary slightly to dilate the orifice. The infant was thin, feeble, and badly nourished, being brought up by hand. The abnormal outlet was small in size and close to the vagina, a slight septum only intervening. The anus was imperforate, but at its site there was a firm, oval elevation, with an indistinct depression in the centre. On examination with a bent probe introduced at the outlet, it seemed that the lower part of the rectum made a bend forwards from the sacrum, the convex part passing close to the integuments where the anus should open. Though the abnormal aperture was obviously insufficient, I did not like to interfere whilst the infant was in such feeble health, but I requested to be sent for in case of any retention taking place. In a few weeks the infant improved in health, but temporary obstructions occurred, and in the beginning of January, 1858, it became evident that the lower bowel was much loaded. There was much straining to pass the motions, and considerable prominence in the perineum; and although there was no actual retention, it was considered undesirable to delay an operation. A narrow, oval portion of skin only was dissected from the site of the anus, and the bowel was reached at the depth of only an eighth of an inch. Directly it was opened, solid evacuations were forced out, and a large quantity escaped. The opening was enlarged so as to admit an elastic gum tube, cut from a No. 9 catheter, which was secured in the passage by elastic tape. The child went on well after the operation. The tube or a plug was passed daily, and kept in for a short time to prevent contraction. In a month the new anus was quite cicatrized. The motions passed by both apertures, but came away of larger size and more copiously from the artificial one. The child thrived well, and was brought to me in the summer of 1862. I found the retentive powers of the anus satisfactory, and on passing my finger it was closely grasped by the sphincter muscle. The perineal aperture was, however, a source of great annoyance from the frequent escape of fæces through it, though none passed the anus except at regular stools. In October, 1862, I pared the edges of the perineal opening and closed it by two metallic sutures. The union was not complete. The droppings of fæces ceased for a time, but defecation continued difficult owing to an insufficient vent. I lost sight of the child until January, 1865, when she was again brought to me, looking in good health and well grown. The anus which I had made would admit only the little finger with difficulty, and fæces still passed by the perineal aperture. There was much straining and frequent defecation. Being satisfied that there was not a sufficient passage, I divided the septum between the two apertures. The result was very satisfactory. The anus admitted the fore-finger easily, but did not seem larger than was necessary. Frequent defecation quite ceased, and the child got into regular habits, and rarely soiled herself at inconvenient times.

A successful operation of this kind was performed by M. Guillon upon a girl aged fifteen, who had, in addition to an imperforate anus with an abnormal opening in the fossa navicularis, a large peduncular fatty tumour attached to the perineum at the site of the natural anus. After excision of this tumour, and the formation of an anus at the part from which it was removed, the opening was maintained by a gum-elastic canula. At the end of fifteen days M. Guillon closed the abnormal opening at the vulva by making two flaps, and uniting them by sutures in the median line. When seen ten years afterwards she had given birth to a child without any accident[*].

Having had experience in both modes of remedying the sixth form of imperfection, I should in future give the preference to the first, viz. the enlargement of the abnormal outlet, in all instances in which the aperture is sufficiently near the site of the natural anus. This is generally the case in the female, and

[*] Bulletin Générale de Thérapeutique, t. xxxiii. p. 477.

sometimes in the male. But in those cases in which a sinus opens in the scrotal raphé or beneath the penis, an **anus must** be made in **the** perineal region, a small **director or** probe, introduced at the abnormal **aperture, being passed** into **the** bowel to serve as a guide to the operator.

7. *Narrowness of the anus.*—In this form, an aperture exists at the proper site, but one too small for the functions of life. The degree of narrowness varies. In some cases, a small-sized probe only can be passed through the anus. The fæces are voided with difficulty and straining, which increase as the motions become solid. There is usually thickening and induration of the integuments around, resulting from inflammation during intra-uterine life, the cause probably of this malformation. It is a rare form, and I have seen only one case of it, the following :—

Case 43.—A female infant, five weeks old, was brought to the London Hospital in December, 1862, with so narrow an anus that only a No. 2 bougie would pass. There was a thickening and bulging of the integuments, especially at the back part. My colleague, Mr. Couper, enlarged the anus by a free incision towards the coccyx, and then connected the mucous membrane of the bowel with the skin with five silk sutures. Bougies were introduced occasionally, but no tendency to contraction ensued, and a free passage was permanently established without difficulty.

Bodenhamer met with two **cases** of this form: one he treated by two lateral incisions, and the other by simple dilatation, with satisfactory results. In all cases **in which** the anus is thickened, as well **as** diminutive, I recommend the treatment by incision as practised in the case just related.

In cases of imperforate anus in which a passage is successfully established, the retentive functions of the bowel generally exist in sufficient force. We have satisfactory evidence on this important point in many of the cases on record; and the existence of an external sphincter, at the natural site of the anus, has been frequently recognized in dissection, in the form of a band of parallel fibres, situated in the median line, without any central separation. In Case 42, in which I successfully established an anus at the natural site by operation, my finger, when introduced four years afterwards, was tightly grasped by the sphincter, and the retentive power of the muscle was complete. As in the imperforate anus the external sphincter consists of a single straight muscle, the surgeon, in operating, should be careful to make his incision exactly in the median line, so that the fibres may be separated equally, and the new anus be brought fully under the influence of the muscle.

In cases of imperforation, unremedied by operation, death is sometimes caused by extreme distension and rupture of the colon, or terminal pouch. In a case in which the operator failed to reach the gut, the infant died in eighty-two hours; and at this early period the colon was found ruptured.[9] My old pupil, Mr. Lys, of Bere Regis, sent me the parts, in a case of imperforate anus, unoperated upon, in which rupture of the pouch, and extravasation into the peritoneal cavity, occurred on the fourth day after birth. The accumulation of meconium, and distension of the colon, vary a good deal in infants, as I have often observed in post-mortem examina-

---

[9] Medical Gazette, vol. xlvii. p. 1077.

tions; and operations have been performed, with success, much later than the period of death in the instances just quoted. In Case 38 the infant was five days old, when I opened the rectum, and saved its life. There are, indeed, on record some remarkable cases of vitality under complete obstruction. Dr. Lyell, of Dundee, met with a case in which an infant lived upwards of twelve weeks without any fecal outlet but the mouth.[1] Still there can be no doubt that in many fatal cases on record the operations were undertaken at too late a period to obtain success. The third form, where the anus is well formed, but the rectum imperforate, is very apt to be overlooked by nurses, and even by medical men, and not to be discovered in time to admit of life being saved by surgical treatment. I know of several instances, indeed, in which castor oil has been given to obtain evacuations when the bowel was impervious.

The most common causes of death after operation are peritonitis and diffuse inflammation of the areolar tissue. The former is generally produced by a wound of the serous membrane; the latter, in infants enfeebled by want of rest and nourishment, by the passage of fecal matter through the tissues of the pelvis. These ill results are chiefly due to faulty methods of operating. In many fatal cases a trocar was used, but this is a most unsafe instrument. A thin, bulging septum, at the end of an anal cul-de-sac, with a fluctuating feel, may be perforated in this way without much risk, but even in this case an incision is preferable. An operation for imperforate

---

[1] Edinb. Monthly Journal, 1852.

rectum should be conducted with the same caution and the same care as an operation for strangulated hernia; and the plunging of a trocar or a bistoury into the depths of the pelvis, in the hope that it may by chance penetrate the rectum, but at the risk of wounding the peritoneum, the bladder, or other important parts, is a rash proceeding, condemned both by reason and experience.

In some instances, troubles in defecation have continued after a sufficient passage for the fæces has been fully established. This is owing to an organic change in the bowel, consequent upon an obstruction of long continuance, subsisting after the removal of the cause. In imperforation the rectum becomes more or less distended, sometimes in a remarkable degree: yet if an outlet of sufficient size be obtained early, the dilatation subsides, and the bowel recovers its natural size and tone. But when the passage or aperture is too small, or subject to recurring contraction, the bowel undergoes changes analogous to those observed in ordinary stricture of this part. A child with an imperforate anus, the rectum opening into the vagina, at the age of four years was successfully operated on at the London Hospital. She died a month afterwards from diphtheria. The rectum was examined, and found to measure five and three-quarter inches in circumference in its most dilated part. Its longitudinal muscular coat was red, greatly developed, and columnar, like the fleshy columns of a hypertrophied bladder. The circular muscular coat was also red and highly developed. The mucous coat was covered with a thick, tenacious mucus, and studded closely with the openings of enlarged follicular glands. A case operated on by Mr. Lane[2] furnishes a remarkable example of excessive distention and hypertrophy of the bowel consequent upon long-continued obstruction. The **infant was** born with an imperforate anus and a small recto-vaginal fistula, through which the **fæces passed until** the **age of four** years and a half, **when the bowel was opened at** the anus, and **the septum between the two apertures** was divided. Death **ensued in twenty-three days. It is** stated that **"the continual strain upon the rectum had** distended it to a size of which it **is hardly possible** to give an adequate description." The rectum and sigmoid flexure of the colon formed an immense **reservoir** capable of containing five pints of fluid, **which** occupied the **pelvis, the** hypogastric, both **iliac, and part** of the umbilical regions, and pushed **the viscera upwards, diminishing** the cavity of the thorax.

When **the vent for the fæces has** long remained insufficient, **and the** bowel has undergone the changes above described, its expulsive functions become seriously impaired and weakened, and the infant consequently suffers in the same way as adults labouring **under stricture of the rectum.** The patient is subject **to great accumulation of fæces, and the fecal collection often sets up mucous** irritation, which **ends in** diarrhœa. **Mr. Le Gros Clark has given me an account of** the state of the patient to whose case I have alluded at page 207, which well illustrates the symptoms in this condition of the bowel. Great difficulty was experienced **for** some time in keeping **the** artificial passage free, and three operations were

[2] Brit. Med. Journal, 1858.

performed to relieve obstruction; but at nine years of age, though this difficulty had ceased, there was a very serious one of a different character. He says, "As far as I can judge by examination with the finger, the cul-de-sac, or sacculated extremity of the bowel above the artificial passage, has become gradually more distended under the pressure to which it has been subjected by the accumulation of feculent matter therein. Long intervals elapse without any attempt at relief, and, finally, a crisis arrives after a fortnight, or even longer, when purging carries off the contents of the bowel, the quantity, of course, being very considerable."

CHAPTER XXII.

COLOTOMY IN CONGENITAL IMPERFECTIONS OF THE ANUS AND RECTUM.

In those cases of imperforation in which the surgeon has operated in the perineum, and unfortunately failed to reach the bowel, life may still be preserved by cutting down on the colon either in the loin or in the groin, and establishing an anus in one of these regions. In order to arrive at a just conclusion as to the advisableness of colotomy and the best method of performing it, the lumbar and inguinal operations must be studied and compared in reference chiefly to three questions:—1. The difficulties of the operation. 2. Its dangers. 3. The condition of the artificial anus, and the inconveniences attending it.

1. The operation of making an artificial anus is

admitted, under ordinary circumstances, to be one of greater difficulty in the loin than in the groin. This arises from the greater depth at which the colon is situated in the lumbar region, and the difficulty of distinguishing the bowel when it is exposed. But there are several circumstances which, in infants, add greatly to the operator's difficulties. In a great fat child the depth is so considerable, that a free incision is required to reach the colon. The kidney varies a good deal in size at this period of life, and when large it overlaps and conceals the bowel. Again, the colon, instead of being distended with meconium, as might be expected, is sometimes contracted, and very hard to find. There are, besides, irregularities in the position of the colon which render it impossible to open the bowel in the left loin without wounding the peritoneum ;[1] and there are some which might prevent the operator from finding the colon in the left groin. With the view of gaining information on these important points, I practised both these operations on the bodies of twenty infants, eighteen still-born, and two deceased a few days after birth; and I afterwards examined the position and course of the colon in each subject. In eighteen of the twenty subjects I found colotomy in the left groin, whether the bowel was distended or not, an easy operation. In one of them the ovary and Fallopian tube protruded, but they were readily put aside and the colon

---

[1] Mr. Duncalfe, of West Bromwich, attempted the operation of lumbar colotomy in a case of imperforate anus, after an unsuccessful search for the bowel at the natural site. He failed also in finding the colon, and closed the wound without giving relief. (Br. Med. Journal, January, 1863.)

reached. In two instances I was unable to find the colon. Both the subjects were well-formed infants, still-born at the full term, one male, the other female. On opening the bodies I observed that the descending colon before reaching the pelvis made rather a sharp curve, and, passing across the abdomen in front of the fourth lumbar vertebra, formed an ample convolution on the right side before terminating in the rectum. Colotomy in the left groin was consequently impossible, in both these cases, owing to the unusual course taken by the descending colon; but the bowel was easily opened in the right groin. This disposition of the colon has been observed in a case of imperforate anus, one already referred to, where death resulted from rupture of the terminal sac. Mr. Lys, who communicated the particulars of the case to me, states, that the descending colon, instead of being directed to the left iliac fossa to form the sigmoid flexure, passed transversely across the spine over the fourth lumbar vertebra to the right sacro-iliac symphysis, and, descending into the pelvis, terminated in a cul-de-sac at the base of the bladder.

In eight of the twenty subjects the colon was readily found, and opened in the loin without wounding the peritoneum. In six the operation was more or less difficult, owing, in two, to the great depth of the gut, in two to its being empty and contracted as well as deeply seated, and in two to a large kidney being in the way and concealing the bowel. Had the subjects been living, I apprehend that the difficulties of the operation would have been increased. In six subjects, lumbar colotomy was impossible without opening the peritoneum, in consequence of the colon being attached by a distinct mesentery, and being loose in the abdomen.[4] In three of the instances, I measured the meso-colon, and found it an inch in width. This serious impediment once occurred to me in performing the operation of lumbar colotomy in the living subject, in a case of imperforate rectum, with the additional difficulty that the colon was empty and contracted, as well as loose in the abdominal cavity.[5] Although deviations from the usual disposition of the colon may prevent or mar the success of colotomy in both regions on the left side, it would appear that this impediment is much more likely to be met with in an operation in the loin than in one in the groin; and if we add to this the other hindrances above mentioned, the difficulties liable to be encountered in the lumbar operation are certainly much greater and more serious than in the inguinal. A surgeon of common skill would find the latter an easy operation.

2. In respect to the dangers of the operation, the cases in which colotomy has been performed are too few in number to admit of a satisfactory comparison between the two methods. In the table in my paper in the Medico-Chirurgical Transactions, the results of these operations, so far as they warrant any conclusion, are much in favour of colotomy in the groin.

[4] M. Lobligeois states (Thèse de l'Oblitération congénitale des Intestins, p. 73), that of eleven infants examined by M. Gosselin in order to determine the arrangement of the peritoneum, in two, the surgeon would have been obliged to divide the membrane in order to open the lumbar colon.

[5] In a case of anus opening into a cul-de-sac, in which Mr. Erichsen was unable to reach the rectum, he performed lumbar colotomy, but was obliged to divide the peritoneum owing to a long floating meso-colon. The infant died of peritonitis. (Br. Med. Journ. 1867.)

In opening the colon in the loin, though the abdominal cavity be unopened, the **wound** required is of greater extent and depth than in inguinal colotomy; and since, as I have already shown, the peritoneum **is very liable to be** wounded in consequence of the **bowel being** only loosely attached by meso-colon, or **being empty and contracted, we** should anticipate **quite as much,** if not greater, risk to life from this operation than from the inguinal. It must be borne in mind, that the fatality attending many of these operations, both in the loin and in the groin, is due in a great measure **to the injuries** inflicted in **the** attempts previously made to establish an anus in the perineum. In an infant with an anus opening **into** a cul-de-sac, upon whom I performed colotomy in the loin, without wounding the peritoneum, death occurred eighteen hours after the operation. A deep puncture had been made through the parts at the end of the sac without reaching the bowel, before the **child** was brought to **me.** On examination of the body, I found that the **colon terminated** in a blind pouch in **the left iliac fossa.** The instrument introduced at the **anus had** penetrated the vagina, passed behind the uterus, and wounded the peritoneum.

3. The sufficiency of an anus, made either in the loin or in the groin, for its necessary functions, as well **as its** condition in relation to the comfort of **the patient, are very important** considerations, not **only as bearing on the comparative value of** the **lumbar and** inguinal operations, **but also** in reference **to the** question **which** parents have to consider, **whether** life is worth preserving with such an infirmity **attaching to it.** After the operation for artificial anus in the groin, the outlet seldom evinces

any disposition to contract, and proves adequate for the passage of **the fæces.** M. Rochard, a naval surgeon and professor at **Brest,** in an interesting memoir communicated to the Imperial Academy of Medicine at Paris,[6] has given an account of the **condition of** the anus in the groin in several patients **who had un**dergone the operation many years previously. He mentions the following cases:—**A robust woman employed in hard labour, operated on in 1813** by Serrand,[7] had excellent **digestion, and** passed solid stools periodically. **When defecation** is about to take place, she is warned by **a sense of** inconvenience and fulness in the left flank. She then withdraws the bandage and compress from her **body, and** replaces it when the desire is satisfied. **No fecal matter passes in the interval,** but a little moisture escapes at times from the upper part of the aperture. **It is only when diarrhœa occurs** that **she is annoyed by the discharge** of feculent matter. —A lady, **operated** on in 1816 by Miriel, constantly enjoys the best health, goes into society, and attends balls, and no one would suspect her to be the subject of any infirmity. She is married, has borne four children, and her pregnancies and labours have been quite normal. She never experiences the least pain in the part. In all the patients observed by M. Rochard, an inversion or prolapsus of the intestine, varying from **about one to** four inches, had taken place; and what is remarkable, the prolapsus occurred exclusively **from the** portion of the bowel beyond **the artificial opening,** owing, probably, to

---

[6] Tom. xxiii.
[7] Vide Amussat sur la Possibilité d'établir un Anus Artificiel.

the circumstance that the colon above was fixed in the loin, whilst the part below **was free** and moveable. The prolapsus did not cause serious inconvenience, and might be almost entirely prevented by a well-fitting apparatus, making a certain amount of pressure.

Colotomy in the loin, to relieve obstructions in **the rectum,** has been performed so often and so successfully in the adult, that we can form a tolerably correct opinion of the convenience of an anus in this region;[3] **and,** through **the** kindness **of** Mr. Walter Bryant, I have had several opportunities of seeing a boy, eight years of age, whose life was saved by this operation. The case is one of much interest.

Imperforate anus, the rectum opening into the urethra. Operation at the anus without result. Successful colotomy in the loin.

Case 44.—The boy was born in South America in January, 1852, with an imperforate anus, the rectum communicating with the urethra. A German surgeon made an incision at the anus to open the bowel, but without success. He consequently performed colotomy in the left loin. The boy thrived, and in the early part of 1860 was brought to England. Mr. Erichsen was consulted as to the propriety of an attempt being again made to establish an anus at the natural site, the operation having been recommended; but he declined to sanction it. I saw the boy with Mr. Bryant in the month of June, 1860. He was healthy, well developed,

---

[3] In eleven of my adult patients who recovered and survived some time after lumbar colotomy, no contraction of the artificial anus ensued, and defecation took place readily. By drawing the bowel to the surface, making a free opening into it, and attaching it to the outer wound, tendency to contraction is avoided, both in lumbar and inguinal colotomy.

active, and remarkably intelligent. The site of the anus was almost natural in appearance, presenting the puckering produced by an external sphincter, but it was closed by skin. There was a cicatrix at the side, about two lines from the median line. The father informed us that the fæces escaped freely from the artificial anus, and had always done so since the operation. There had been no tendency to contraction. The boy had no sensation or warning before an action took place, which occurred usually at night. He was seldom inconvenienced by feculent discharges at other times, unless the bowels were relaxed. His chief trouble was painful and difficult micturition. This occurred at uncertain times, but had increased in frequency during the last two years, and he usually suffered from it two or three times during the week. It obliged him to strain violently in passing water, and caused altogether great distress. There was an inguinal rupture on the right side, which seemed to have resulted from the violent straining. The admixture of feculent matter with the urine passed on these occasions showed that some portion of the fæces escaped occasionally into the colon beyond the artificial anus, and from thence into the urethra, obstructing the passage. He was obliged to be very careful in his diet, and to avoid swallowing currants and other indigestible matters. The anus was of ample size, and oval in shape, with the red mucous membrane constantly protruding in close-set folds. The surface of the membrane was not at all sensitive. The finger passed easily into the bowel, both above and below, and was slightly girt by the actions of the abdominal muscles around the aperture. The surrounding skin was free from irritation. A prolapsus of the bowel readily took place on the slightest straining, and usually occurred when the part was uncovered for the purpose of cleanliness. The protrusion was easily reduced by slight pressure; but on one occasion, some months previously, a very considerable inversion occurred, and it was so difficult to replace that chloroform was administered to effect the reduction.* The position of

* M. Larrey showed, at a meeting of the Société de Chirurgie de Paris, an infant seven months and a half old, born with an

the anus at the outer part of the loin, three inches from the spine, did not render personal attention to it difficult or inconvenient, and the little fellow could easily replace the ordinary protrusion by slight pressure with a napkin or sponge, whilst a bandage giving gentle support was sufficient to restrain it at other times. He ran about and took exercise like other boys, and no one would suppose him to be labouring under any infirmity, until obstruction occurred in micturition. To prevent the occurrence of this very serious evil, two modes of proceeding seemed to offer. 1. The constant wearing of a plug in the lower portion of the colon to prevent the entry of fæces into it. 2. An operation for closing the lower opening into the colon. As the second plan would not be free from some risk, we agreed to make trial of the first; and a piece of sponge, of proper shape, secured with a strong ligature, was lodged in the colon below the outer opening. This seemed partly to answer the purpose, and to save the boy distress in micturition; but it did not entirely prevent the entrance of liquid fecal matter, and the plan was not persevered in by the parents. The lower portion of the colon was, however, washed out daily by injections.

In a case in which the fæces ceased entirely to pass beyond the artificial anus we should expect more or less wasting of the bowel beyond. Dr. Harris has recorded an interesting dissection of the parts from a patient who died of disease of the kidneys five years and a half after lumbar colotomy for vesico-intestinal fistula performed by Mr. Hakes,

anal imperforation. M. Maisonneuve, having failed to reach the bowel in an operation in the perineum, performed colotomy in the loin with success. The anus was the seat of an extensive double prolapsus, and had a remarkable appearance, the ends of the inverted bowel diverging in opposite directions. No attempt had been made to reduce it. Bulletin de la Société, tom. vi. p. 410.

of Liverpool. The intestine between the artificial anus and the cicatrix of the recto-vesical opening was completely atrophied, reduced to a fibro-areolar cord.[1]

I am indebted to the courtesy of Dr. A. Amussat for some account of a case in which his father performed lumbar colotomy, in 1852, on a boy with an anus opening into a cul-de sac. The boy was living at the age of eight in good health, performing his functions regularly. A wax bougie was retained in the artificial anus by a belt. Amussat had only two successful cases of lumbar colotomy in infants. The first one died at the age of seven.

In health, an artificial anus is productive of much less personal annoyance than is commonly supposed. Some degree of protrusion of the mucous membrane of the colon will always be liable to occur in consequence of the deficiency in the abdominal wall, and the want of support at the part. In a sound condition of the digestive functions, the fecal evacuation is regular and periodic, and the anus is almost free from discharge at other times, whilst a well-fitting apparatus is sufficient to restrain protrusion, as well as to prevent unpleasant oozing from the opening. When diarrhœa occurs, or unhealthy gases are generated, then the annoyances are considerable, and may render the patient unfitted for society. It has been argued, that an anus in the loin is much more inconvenient than one situated in the groin, as regards personal attention, whilst one in the groin is likely to be more repulsive in the relations of

---

[1] Liverpool and Manchester Medical and Surgical Reports, 1875, p. 104.

adult life. I confess, that I see very little to justify a preference for either operation, on the ground of the position of the anus; but the greater difficulties and dangers of lumbar colotomy in infants, which I have already pointed out, would certainly induce me to give the preference to the inguinal operation for the preservation of life in cases of atresia. It was performed in the following case:—

Imperforate anus and rectum. Colotomy in the groin.

Case 45.—A female infant, born at the seventh month, with an imperforate anus, was brought to the London Hospital on the twelfth day after birth. Mr. Gowlland made a persevering attempt to reach the bowel in the perineum without success. I saw the case with him, and inguinal colotomy was resolved on, with only a faint hope of success under the circumstances. The colon was reached in the left groin with the greatest ease, and it was attached to the abdominal walls before an opening was made into it, so that no meconium could escape into the peritoneum. The infant cried lustily during both operations, and was much less exhausted by them than might have been expected. She survived six days.[2]

In a recent discussion at the Imperial Academy of Medicine in Paris,[3] the question has been raised, whether the left side should be selected for the in-

---

[2] I have since performed inguinal colotomy on the left side in two other cases of imperforation. The circumstances were unfavourable. There was no difficulty in the operation, which was followed by a free discharge of meconium. Both infants died in a few days. M. Guersant opened the colon in the groin eleven times in succession, without saving one of his patients. (Bull. de Thér. tom. xlix.)

Bulletin, tom. xxiv. 1858, 1859, p. 445.

guinal operation. M. Huguier strongly insisted on the importance of performing colotomy in the right groin. He stated, that during intra-uterine life the sigmoid flexure is enormously developed, and finding itself restricted in the left iliac fossa, passes over to the right iliac fossa, and thence dips into the pelvis to join the rectum. This disposition is observed in children up to the age of eighteen months or two years. The opening, therefore, if made in the bowel on the right side, would be nearer its termination, and the infant would be less liable to prolapsus, in consequence of the retraction of the sigmoid flexure, as the infant grows. I practised colotomy on the right side in five infant subjects. In three the colon was easily reached, and, in two of these, the convolution of the colon, from the left groin to the right, measured nine inches in length. In a fourth, when the abdomen was opened, small intestine presented, and no colon could be found; for it appeared that the bowel ascended from the left iliac fossa, and making a considerable convolution in the left groin, five inches in length, terminated in the rectum. In a fifth subject the cæcum appeared, and no colon could be reached. The large intestine, on leaving the left iliac fossa, made a convolution upwards, three inches in length, but did not approach the right groin. M. Boucart also examined a large number of bodies of infants new-born, or some days old, with reference to the question raised by Huguier. Boucart found the transverse disposition of the colon quite exceptional. In 144 instances out of 150 the sigmoid flexure was in direct relation with the abdominal walls on the left side. In operations practised on the right side the cæcum generally presented itself, the sigmoid **flexure** being very rarely met with.⁴

We may infer from these few examinations, **that the disposition of the colon in** infants, described **by Huguier, is not so constant as** he states. **The** colon, **at the early** period of life, is largely developed, and **forms ample** convolutions, after reaching the **left** iliac fossa, varying in extent, however, a good **deal** in different subjects; and, though its usual course is directed to the right groin, the exceptions are too numerous to render the **inguinal** operation on the right **side as feasible as on the** left. **But little importance can** be attached to the circumstance **of** an artificial anus on the right side being somewhat nearer the termination of the alimentary canal; and if it seem probable that, after an operation on this **side,** a prolapsus is less likely to occur from the lower **part of the** bowel, we might equally expect a greater liability to inversion in the upper portion which would be free **and** loose **in** the abdomen. The only **case,** in which, so far as I know, colotomy **has been performed in the right groin is related by Mr. Bryant.**⁵ The colon was opened without difficulty, and secured to the integument by two thick ligatures; but the result was unfortunate, and will **serve** as a warning against a repetition of the operation. **Within a few** hours, during a fit of coughing, **the stitches gave** way, and small intestine **protruded, the** child dying ten **hours** after the operation. There **must have** been some great stress, some dragging away **of the** colon to cause the

⁴ **Archives Générales** de Méd., Nov., **1863,** p. 621.
⁵ Surgical Diseases of Children, p. 40.

stitches to give way thus early, and to leave the space free for the escape of the small intestines.

In performing inguinal colotomy on infants, an oblique incision from an inch and a half to two inches in length should be made in the left iliac region above Poupart's ligament, reaching a little above the antero-superior spinous process of the ilium. The fibres of the abdominal muscles should be divided on a director passed beneath them, and the peritoneum should next be cautiously opened to a sufficient extent. The colon would most likely protrude, but if small intestine appear, the colon must be sought for higher up. A curved needle armed with a silk ligature should be passed lengthways through the coats of the upper part of the colon, and another inserted in the same way below, and the bowel, being drawn forwards, should then be opened by a longitudinal incision. The colon must afterwards be attached to the skin forming the margin of the wound, by four sutures, at the points of entry and exit of the needles. I prefer silk ligatures to metallic, as the former are less liable to cut or ulcerate through the delicate tissues.

THE END.

INDEX.

ABSCESS near the anus, 88; **treatment** of, 93.
Amussat on excision of coccyx **in imperforate** anus, 203; case of anus opening into *cul-de-sac*, 204; his operation in deficiency of rectum, 206; successful case of lumbar colotomy for imperforation, 236.
Anæsthetics in diseases of the rectum, 4.
Anal tumours, 188.
Annesley, Sir James, on stricture after dysentery, 128.
Anus, chaps and sores at, 14; **treatment** of, 14; congenital imperfections, 196; epithelial cancer **of,** 164; **cases** of excision of, **164;** organic contraction of, 190; contraction of, **after excision of** epithelial cancer, 191; **after healing** of ulcers, 191; **congenital narrowness** of, 222; **sufficiency of artificial,** 231.
Aperients in hæmorrhoids, 45—48.
Atresia, 196.
Avery, case of syphilitic ulceration of rectum, 112.

Bodenhamer, cases of congenital narrowness of anus, 222.
Boucart, on disposition of colon in infants, 238.
Bougies, of elastic gum, 137; of compressed sponge, 138; of laminaria digitata, 138; action in curing stricture, 145.
Bouisson, cases of malformations **of** rectum, 198.
Boyer, operation **for irritable ulcer, 10.**
Brodie, **on** situation of inner orifice of a fistula, 91; **on** mode of operating in fistula, 99; case of rectovaginal fistula cured by operation, 108; excrescences in stricture, 134.
Bryant, operation of colotomy by oblique incision, 181; case of colotomy in right groin, 239.
Bush, on cure of stricture, 146.

Cancer of rectum, 154; seat of, 155; symptoms of, 156; pregnancy in, 159; relative frequency of, in the sexes, 159; treatment of, 159; excision of, 161; lumbar colotomy in, 162.
Catarrh of rectum, 109.
Chassaignac, removal of piles with the ecraseur, 59.
Clark, Dr. Andrew, on minute stricture of villous tumour, 85.
Clark, Le Gros, operation in **case of** imperfect rectum, 207.
Colles, on cure of stricture, 146.
Colon, rupture of, in imperforate anus, 223; author's investigations into the position of the colon in infants, 229, 238.
Colotomy, diseases requiring it, 178; in intractable ulcers of rectum, 153; in cancer, 162; operation described, 179; Bryant's operation by oblique **incision,** 181; difficulties in colotomy, 181; author's cases, 184; **in congenital** imperfections, 227; **lumbar** and inguinal compared, **227;** inguinal **operation** on infants described, 240.
Contraction of anus after removal of piles, 59.
Cooper, Sir Astley, case of bleeding after excision of polypus, 80.
Copland, on incision in irritable ulcer, 11.

Dieffenbach, excision of carcinomatous rectum, 161; operation in imperfections of the rectum, 206, 212.
Dittel, Professor, treatment of fistula by india-rubber cord, 101.
Dupuytren, on cure of stricture, 146.

Epithelioma of anus and rectum, 164; excision of, and dangers of hæmorrhage, 170.

Fissure of the anus, 6.

R

INDEX.

Fistula in ano, 86; mode of origin, 88; in phthisical subjects, 91; complicated, 92; symptoms of, 93; mode of examining, 94; **cure by** cutting operation, 95; after treatment of, 97; mode of dividing sinus high up the bowel, 99; **hæmorrhage** after operating for, 100; **treatment of** by ligature, 101; **treatment of blind external**, 103; **treatment of blind internal**, 104; liability to be overlooked, 104; treatment of complicated, 105; **propriety of operating in phthisical patients**, 107; connected with carious bone, 107; congenital fecal, case of, 197.

Galvanic cautery in cure of **hæmorrhoids**, 54; of fistula, 102.
Giraldés, cases of irritable ulcer treated by forcible dilatation, 14; case of punctured bladder in operation for deficient rectum, 204; in cure of hæmorrhoids, 54.
Godard, curious case of **anus opening** into cul-de-sac, 204.
Gosselin, on syphilitic **ulceration of** the rectum, 113; **on excrescences in** stricture, 134.
Gowlland, Mr., cases **of double division** of sphincter muscle, 18.
Goyraud, operation for imperforate anus with rectum opening externally in an abnormal situation, 216.
Groin, colotomy in, 228; for imperforation, cases of, 232, 237; condition of artificial **anus in**, 232.
Guersant, his treatment of **prolapsus by actual cautery**, 75.
Guillon, successful **operation for imperforate** anus, **rectum opening into vagina**, 221.

Harris, Dr., dissection of bowel five years after lumbar colotomy, 235.
Hæmorrhage from piles, 40; periodic, 41; venous, 42; arterial, 42; mode of arresting, 60; case of obstinate, 61; in prolapsus, 71; after excision for cure of prolapsus, 76.
Hæmorrhoidal veins, arrangement of, 29.
Hæmorrhoids, external, 29; internal, 30; changes of structure in, 30; error in the description of piles as occurring in two rows, 31; appearance of internal hæmorrhoids when protruded, 32; coexistence of external and internal hæmorrhoids, 33; causes of hæmorrhoids, 34; symptoms of external hæmorrhoids, 35; congestion and inflammation of, 35; ulceration of, 35; symptoms of internal hæmorrhoids, 36; connexion between hæmorrhoids and disorders of urinary organs, 37; complication of hæmorrhoids, with enlargement of prostate gland, 38; strangulation of internal hæmorrhoids, 40; **mode of** examining, 43; general treatment of, 44; aperients in hæmorrhoids, 45; treatment of external by incision, 46; by excision, 46; caution against too free excision, 47; treatment of internal, 47; by cauterization, 50; by actual cautery, 53; by ligature, 54; treatment after operation, 57; removal by ecraseur, 59; advantages of cautery and ligature of, compared, 63; operations on, in aged person, 64; mechanical treatment of, 65.
Hey, on cure of prolapsus by excision of hæmorrhoidal excrescences, 74.
Holt, case of extirpation of cancer of rectum, 160.
Huguier, on disposition of **colon in** infants, 238; on colotomy **in right groin**, 238.

Imperfections, congenital, of **anus and rectum**, 196; relative frequency in the **sexes**, 198; causes of, 198; relations of peritoneum in, 200; defective development of pelvis in, 201; causes of death in, 224; troubles in defecation after operation for imperforation, 225.
Injections, cold and astringent, in **piles**, 48.

Laminaria, digitata bougies, 138.
Lane, remarkable case of hypertrophy of bowel after operation for imperforation, 226.

Ligature in treatment of fistula, 101; instruments for passing the ligature in fistula, 101.
Loin, colotomy in, 179; condition of artificial anus in, 233.
Lyell, Dr., case of long duration of life in imperforation, 224.

Martin, Sir Ranald, on rarity of stricture after dysentery, 129.
Mason, Dr., colotomy in stricture, 153.
Mayo, his case of bleeding after too tight ligature of polypus of the rectum, 80; his case of hæmorrhage after division of a stricture, 140.
Melanotic cancer, 174.
Miller, Dr., case of imperforate anus with rectum opening into urethra, 210.
Miriel, case of lady with **artificial** anus in the groin, 232.
Morgagni, case of a woman who lived to one hundred passing all her **fæces** by vagina, 211.

Nélaton, remarks on the removal of piles with the ecraseur, 59.

Obstruction in stricture, 131, **175**; operations for, 176; after operation for imperforation, 226.

Pelvis, contractions of, in **congenital** deficiency of **rectum, 201.**
Pennell's **case of colotomy, 179,** 183.
Perineum, author's **case of rectum** opening into, 217.
Petit, case of operation in imperforate anus, 202.
Piles, attacks of, **39; bleeding piles,** 40.
Polypus, 78; symptoms of **in children,** 79; treatment of, 79; **danger** of bleeding after excision **of, 80;** fibrous, in adults, 82, ulceration of, 83; **removal** of, by **ecraseur,** 83.
Pregnancy, hæmorrhoids in, 34.
Prolapsus, **66**; in children, 68; in adults, **69**; combined with internal piles, **69; strangulation** of, **70; bleeding from, 71**; treatment in **children, 72; treatment in** adults, **74; cure by escharotics, 74;** danger of stricture from application of escharotics for, 75; cure of, by excision, 75; **risks** of bleeding after operation for, **76**; in artificial anus, 232, 236.
Prostate gland, influence of enlargement of, in producing hæmorrhoids, 38.
Prurigo ani, 192; **causes of,** 192; treatment, 194.
Purgatives, influence of in producing hæmorrhoids, 34, 39.
Pyæmia after operations for hæmorrhoids, 63.

Quain, his case of peculiar bleeding tumour of the rectum, 85.

Rectoscopes, 4.
Rectum, complaints **of** mistaken for other affections, 2; **mode of examination in,** 2, 4; irritable, **19;** morbid sensibility of, 22; **neuralgia** of, 25; treatment of, 27; mental affections **of,** 28; purulent discharges from, 109; chronic ulceration of, 110; stricture of, 121; compression of, by neighbouring viscera **and** tumours, 136; cancer of, 154; epithelial cancer of, 164; atony **of,** 184; cause of fæcal accumulations in, 185; congenital imperfections of, 197.
Ribes, **on situation of the** inner orifice **of a fistula in ano,** 90.
Rizzoli, operation for imperforation **with** recto-vaginal opening, 215.
Rochard, on condition of artificial anus in the groin, 232.

Sharpless, Dr., case of operation in imperforate anus, 213.
Simpson, Sir James, on obstruction of rectum by scybalous masses, 187.
Speculums, rectal, 3.
Sphincter muscle, irritable, 15; hypertrophy **of,** 16; occurrence in hysterical females, 16; treatment of, 17; division of the muscle, 16; partial **excision** of, 169; existence of **in imperforate** anus, 223.
Stricture, **121**; pathological changes in, 121; seat of, 123; double, 123; causes of, 124; produced by nitric acid applied for cure of prolapsus, 75, 146; case of double, after ulceration from chronic dysentery, 126;

disease of middle life, 129; symptoms of, 129; mode of examining, 132; excrescences of mucous membrane in, 133; difficulty of detecting when high up, 135, 141; treatment of, by gradual dilatation, 137; by forcible dilatation, 138; by incision, **139; risks of** incision in, **140;** dangers of passing bougies **high up** in, 142; general treatment **of, 143;** cases of cure, 147; treatment of, when inveterate, **by** colotomy, **151.**

Syme, his cases **of bleeding polypus in** adults, 85; **on mode of operating** in fistula, **99.**

Tetanus, cases of, after operations for hæmorrhoids, 63.

Tumours obstructing rectum, **136.**

Ulcer, irritable, 6; sufferings produced by, 7; pedunculated pile in, 7; consequent **on** operations, 9; **more** frequent **in** women, 9; **Boyer's** operation of incision in, 10; treatment before operation, 11; **mode of** performing incision in, 11; treatment **of** after operation, 12; treatment **of** by local **applications** without operation, 13; treatment of by forcible dilatation, 14; Giraldés' **cases** of failure of dilatation, 14.

Ulceration, chronic, pathology of, 110; causes of 111; syphilitic, 112; tubercular, 113; symptoms of, 114; treatment of, 115; cases illustrating treatment, 115; rodent ulcer, 120; case of, 121.

Urethra, cases of **congenital opening of rectum into, 208.**

Urinary organs, **irritation of** in hæmorrhoids, 37.

Uterus, influence of diseases of in producing piles, 38; displacements of, obstructing rectum, 136; case of bifid, with imperforate anus, 214.

Vagina, cases **of** congenital opening of rectum into, 211; double vagina and bifid uterus with imperforate anus, case of, 214.

Verneuil, cases of resection of coccyx in imperforation of the rectum, 203.

Vicq-d-Azyr, his operation for remedying imperfect anus, rectum opening into vagina, 212.

Villous tumour, 84.

Ward's **paste, use of** in hæmorrhoids, 48.

Warts **in** epithelial cancer, **168;** in vicinity of anus, 189.

Woodman, Dr., cases of polypus, **79.**

Wutzer, excision of rectum, 161.

CORRIGENDUM.
Page 54, line 3, *for* Esmarck *read* Esmarch.

www.ingramcontent.com/pod-product-compliance
Lightning Source LLC
Chambersburg PA
CBHW021345230426
43666CB00006B/411